LEARN MATH IN 7 DAYS !

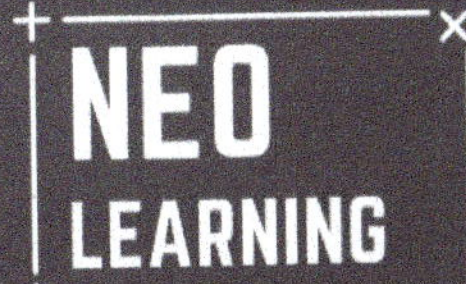

Learn Something New

SPECIAL FEATURES

Brain based learning techniques for memory improvement and meditation

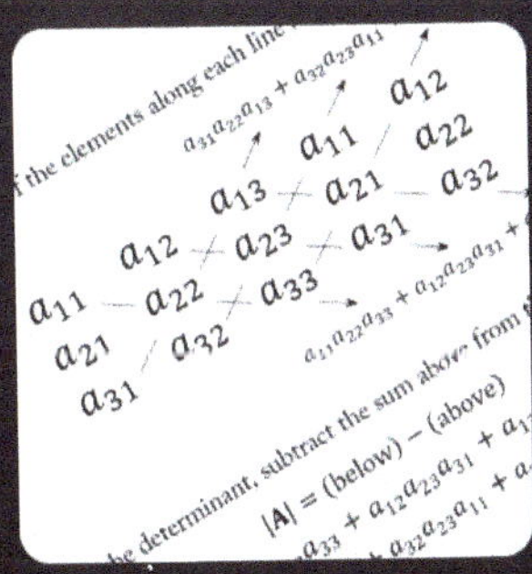

Includes numerous shortcuts, illustrations and easy methods

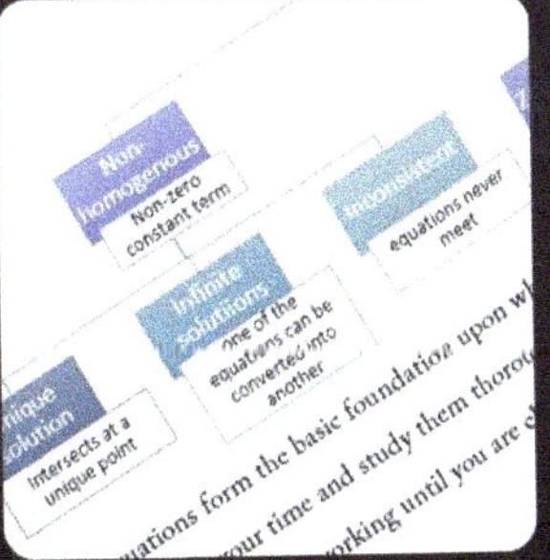

Numerous examples and problems to work out, and a final Quiz for final brush up

Visit: www.neolearningbooks.com for freebies and more..

www.neolearningbooks.com

SEVEN DAY STUDY PLAN

	SESSION 1	SESSION 2	SESSION 3
DAY 1			
DAY 2			
DAY 3			
DAY 4			
DAY 5			
DAY 6			
DAY 7			

Vineeth Remanan

MATRIX AND DETERMINANTS IN ONE WEEK

International Study Version

Learn

Something

New

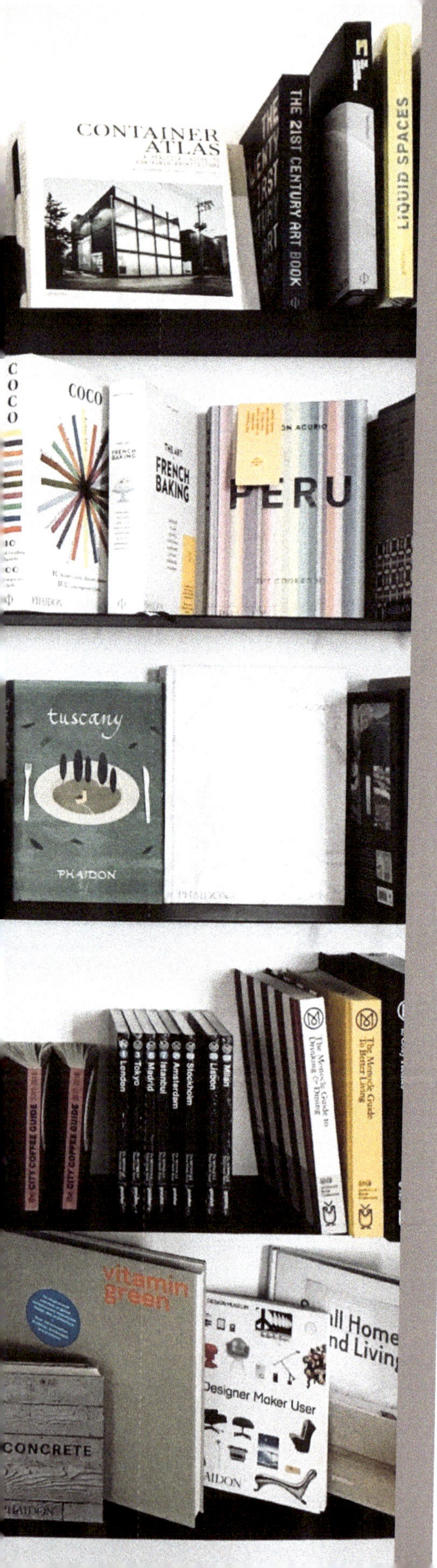

MATRIX AND DETERMINANTS IN ONE WEEK

ISBN: 978-93-5526-610-1

www.neolearningbooks.com

Mathematics – Linear Algebra

Written, edited and designed by
Vineeth Remanan.

Email: learnsomethingnew@neolearningbooks.com

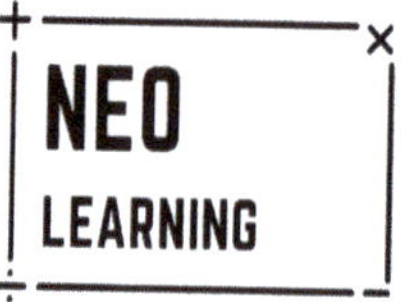

Learn
Something
New

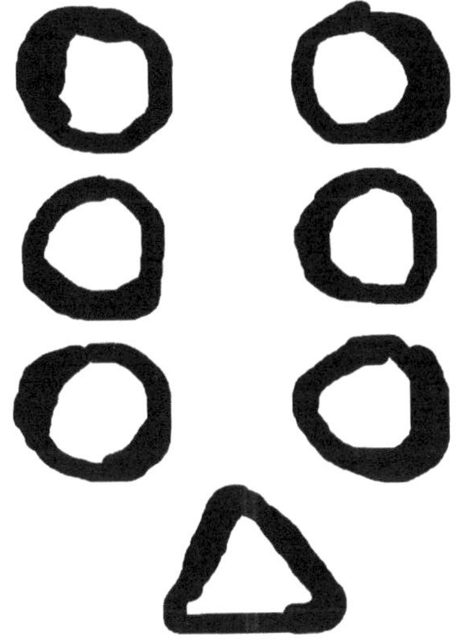

CONTENTS

MONSTER AND HORROR MOVIES
King and Maxwell
LOS SANTOS INOCENTES
COLD MOUNTAIN
MICHAEL CONNELLY
The Black Box
Equinox
LA TIENDA ROJA
Young people's story

現代名作集
現代名作集

金額の見方

How to use this book - A preface on learning

Learning is one's wilful attempt to acquire knowledge at the expense of time and intense focus. It requires the desire to understand, the resolve to push through, the patience to persist and the courage to fail.

No knowledge in this world is incomprehensible, no skill in this world in unlearnable. After all, it was mere mortals like us who invented them.

The only hard part is the path one has to follow. A path of dedication, discipline, consistency and commitment. That is the only prerequisite required to start this course.

About this book

This book is a short yet consistent introduction to the world of matrices. It is divided into seven chapters – "The Seven Days", each consisting of three sessions, making it a total of 21 sessions, to be completed in one week.

But there are some caveats.

Identify yourself

Before starting the course, find out what kind of a learner you are?

Are you a fast learner, or a slow learner? Are you good at math, or having a hard time with it?

Use this book as a one week course, only if you are a fast learner. Otherwise, don't.

- **If you are a fast learner**:

Think of this as an athletic training camp. Athletes and competitive sports persons often participate in organised and rigorously focused training schedules to prepare themselves for upcoming sporting events.

For instance, an intense MMA training camp usually consists of 4-5 hours of training per day, split into 2 or even 3 sessions, for 5-6 days a week.

Constructing a similar schedule involves splitting a day into three sessions of 2 hours each, with enough time gaps to reset. The resetting involves refreshing, snacking, meditating and recalling the topics of the previous sessions.

Read through the topics and work out the problems. Continue to the next session only if you are confident about the current. Even if you are fast, sometimes you may not be able to keep up. Don't worry, that's ok. Take your own time. Forget the time constraints, and focus on learning.

- **If you are a slow learner**:

Appreciate the fact that every individual is unique and different. Thomas Sowell once wrote, "Nobody is equal to anybody. Even the same man is not equal to himself on different days." The same principle also applies to learning.

While fast learners readily solve problems, slow learners acquire a deeper understanding of the subject. They are not fast, because they need to convince themselves the things they are studying. Once they do that, they remember things for a longer period of time. For them, time constraints are pointless.

Unfortunately, the world is designed for quick learners, who often excel in exams and achieve higher GPAs. As a result, the slow learners whom I call the deep learners, are often marginalized and are treated as insignificant. But these are the people who might "cure our cancer or build multimillion-dollar industries".

So, if you feel left out, don't worry. Move at your own pace, take your own time, work hard and push it to the limits. The only thing that matters is the willingness to learn and the belief in yourself.

Principles of Brain-Based Learning (BBL)

The human brain is a sense making machine. Its purpose is to make sense of the information coming in through our senses, and construct various logically coherent and ever-changing structures of data. These data structures are manifested physically in the form of nerve cell connections called synapses.

One cubic millimetre of the cerebral cortex contains roughly 300 million synapses! The total number of connections in the entire brain is literally innumerable. They are responsible for our bodily functions, memory, emotions and our subjective free will.

When we learn a new subject or a skill, the nerve cells realign themselves forming new connections in order to accommodate the new information. This is called neuroplasticity.

But the new connections may not be strong as the existing ones, meaning, they may not conduct information quickly and effectively like the others. But the more we use them, they become stronger, faster, and more efficient. This is why we remember our friend's name, not our friend's friend's.

A new word, a new skill, or a new name is committed to our memory only if we use it often. Otherwise, the connections become weaker and weaker, ultimately resulting in oblivion.

This brings us to the first practice of active learning.

Spaced Repetition

The spacing effect is considered as "one of the most robust effects in psychology". It is an evidence-based learning technique in which the newly formed connections are repeatedly activated with enough time gaps in between.

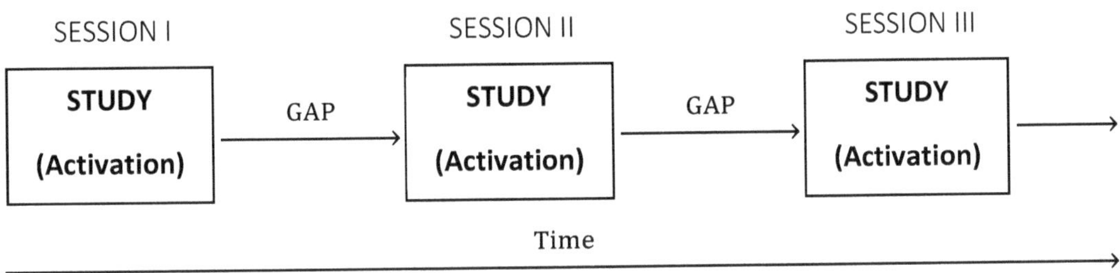

By the repeated activation, our brain recognises the importance of those connections, forcing it to strengthen the associated synapses. But the human body is not capable for a long and continuous expenditure of energy, without some form of rest. So, the gaps between the sessions are also important. A good night sleep, a daytime nap, or even a few moments of quite wakefulness could significantly improve our learning and memory.

So, instead of long, continuous study sessions, break up and distribute your learning throughout your day, or your week, or your month, with enough gaps to rejuvenate.

A few printable resources for planning and scheduling are available on our website for free. See back cover for details.

The Learning Process

The spaced repetition is shown to yield remarkable results through the repeated activation of the synapses in both clinical and academic settings. But a mere statement that: 'Studying with enough time gaps is good' is nothing more than a prescription for a scheduled study routine.

It doesn't answer about the nature of the learning process, or the methods involved in the consolidation of knowledge, neither and effective retrieval strategy to recover the stored information. For that, a further probing into the nature of learning and memory is essential.

Consider the scenario in which we are starting to 'learn something new'. Regarding the high success rate of the spaced repetition, we intend to follow such a routine.

On the first session, don't expect to learn everything. If you were able to, that's ok, good job. If you couldn't learn anything, don't worry. No one is expecting you to become a master in half an hour. Another outcome may arise in which you might tackle some concepts, but some were confusing and unclear.

If you have comprehended everything, start the next session by recalling them, then move on to the next.

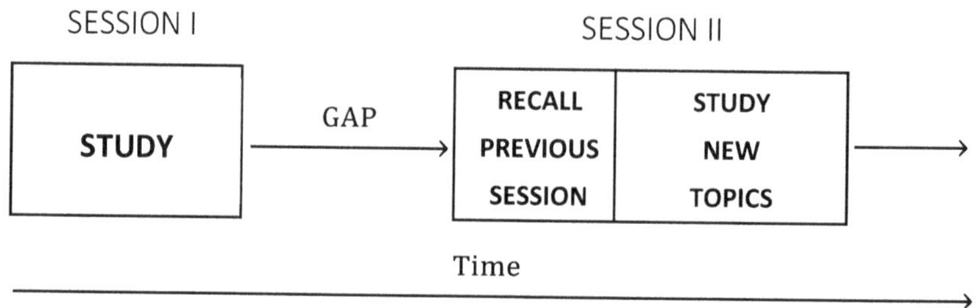

If you couldn't comprehend anything, don't worry. Remember, learning is never easy, even for the greats. So, be patient, study the topics once again and repeat until comprehension.

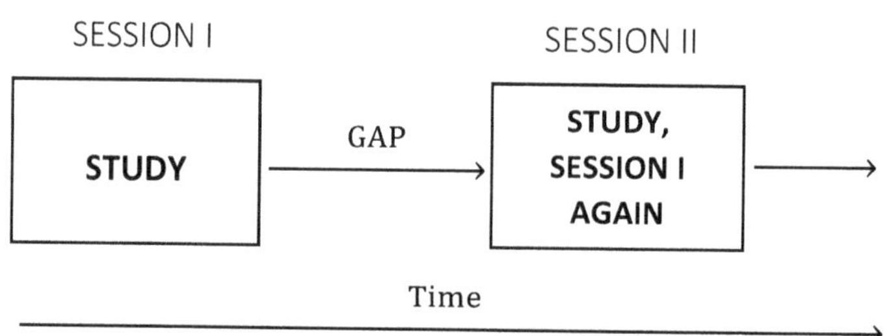

If you find yourself in the third scenario, that is you've tackled some concepts, but some were unclear. Start by recalling the concepts that you've comprehended, then study the topics that were unclear.

For instance, out of the topics A and B, assume you have learned only A. Then on the next session, recall A, then study B.

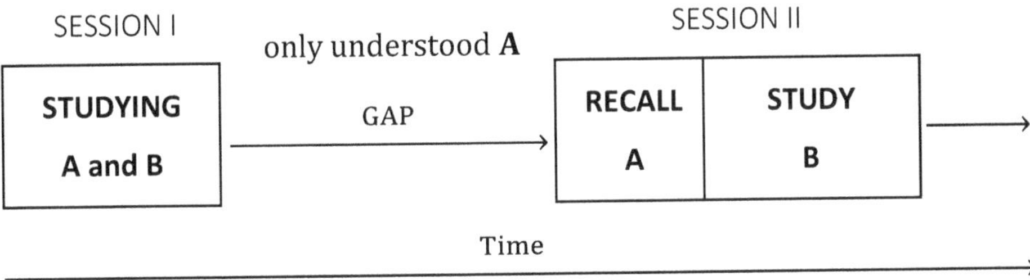

Flash Card Recalling (An Effective Retrieval Strategy)

Even though we have used the word 'recalling' quite a few times, we haven't truly discussed what it actually is.

Recalling is simply pulling information straight out of your brain without study notes or reading materials.

The absence of reading or watching forces our brain to simulate the activity originally created during the learning process. It is the most effective way to consolidate a long-term memory.

An effective execution of this technique can be done using flash cards. They are small, concise paper cards bearing keys to a particular topic. While recalling, the student must read the key and retrieve the answer.

For instance, let's say we have learned that the first president of the United States was George Washington. Then at the end of the session, make the flash card.

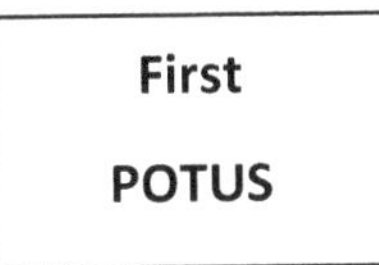

POTUS meaning, President Of The US.

On the next session, instead of reading the notes, just see the card and retrieve the answer.

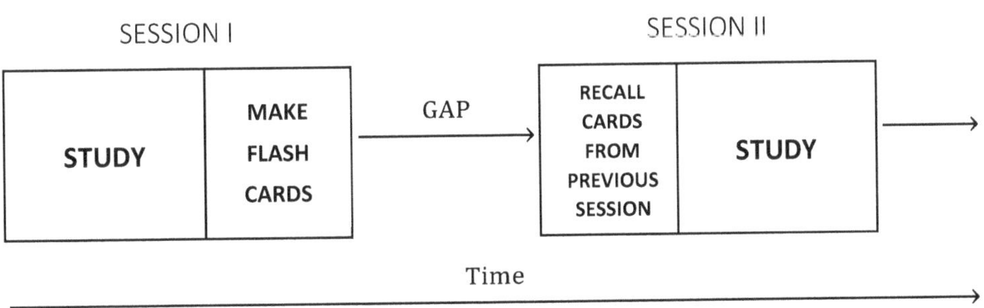

A key can be anything but the answer. It can be a code, a question, an acronym or a phrase. Don't write multiple keys into a single card. A single card should be for a single key, and always recall one at a time.

Focus – The prerequiste for learning

Up until now, all of our discussions were concerned with the process of learning and the methods involved in it. But we haven't quite discussed what its prerequisites are. Of course, an emotional motivation is required for learning. But emotions are also the universal prerequisite for all 'human actions.'

The thing that differentiates learning from almost anything, is that a focused mind is a necessity.

But in a world full of distractions, the ability to focus is almost a rarity. It is a superpower to narrow down our attention to a sole subject. It is also a test of our minds and ourselves to achieve such a state.

Recent studies in the field of learning concludes that slow learning along with some learning disabilities might be connected to the inability to focus. Therefore, it shall be the duty of every learner to prioritize the nurturing of a sound and a focused mindset.

A regular practice of meditation might seem silly, but is shown to improve our focus, attention, perception, memory, language, and even the control over our emotions.

Major corporations like Apple. Google, Nike etc. promote mediation amongst their employees in order to increase their productive output.

The benefits of meditation to mental health and productivity are enormous. It is also an ongoing and an exciting field of contemporary research.

Steps for meditation

i. **Be comfortable:** There is no fixed position or posture for meditation. You can either be on a chair, or on a bed. You can be in your room, or on a bus. As long as you are comfortable, it doesn't matter where you are.

ii. **Breathe in, Breathe out:** Meditation is all about being in the present. At present, what are we doing?... We are breathing.

So, close your eyes, breathe in, breathe out. Don't try.... just feel.

Feel the air rushing in and out. You can either focus on your nose, or on your chest, or on your belly.

iii. **Lost in thoughts:** The moment you start breathing, your mind starts to wander. Various thoughts rush into you, grabbing away your attention, and the next thing you know, you are completely lost in thoughts.

Don't be worried and don't get frustrated. It is our very nature to produce thoughts. The key to meditation is not to avoid them, but to notice them. Notice that your mind has wandered. If you have done that, you have unlocked it.

iv. **The comeback:** After noticing your wandered mind, the next logical step is to grab it, place it on your breath and start over once again.

Meditation is not a linear process, it is cyclical. You focus on your breath, you get distracted, you notice your distraction, then comeback and repeat the process over and over until you are relaxed, calm, confident and ready to learn.

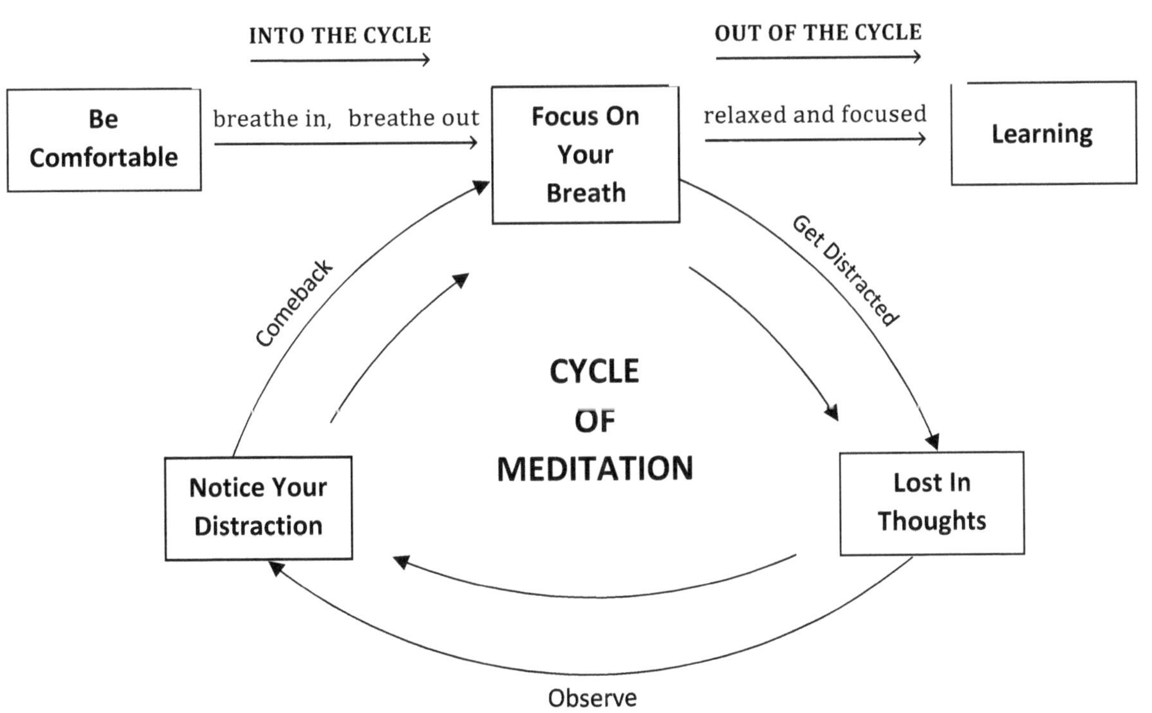

The fulfilment and satisfaction involved in a deep and meaningful learning is almost forgotten in these modern times. True learning is misplaced with cram courses and true satisfaction is misunderstood as GPAs. Even though this book is intended for a short course in mathematics, it shall not deceive the reader in believing that, there is nothing more to it. The reader should see this course as a gateway, a gateway into a wider, richer and a more profound realm of knowledge.

Good luck

Vineeth Remanan

Neo Learning

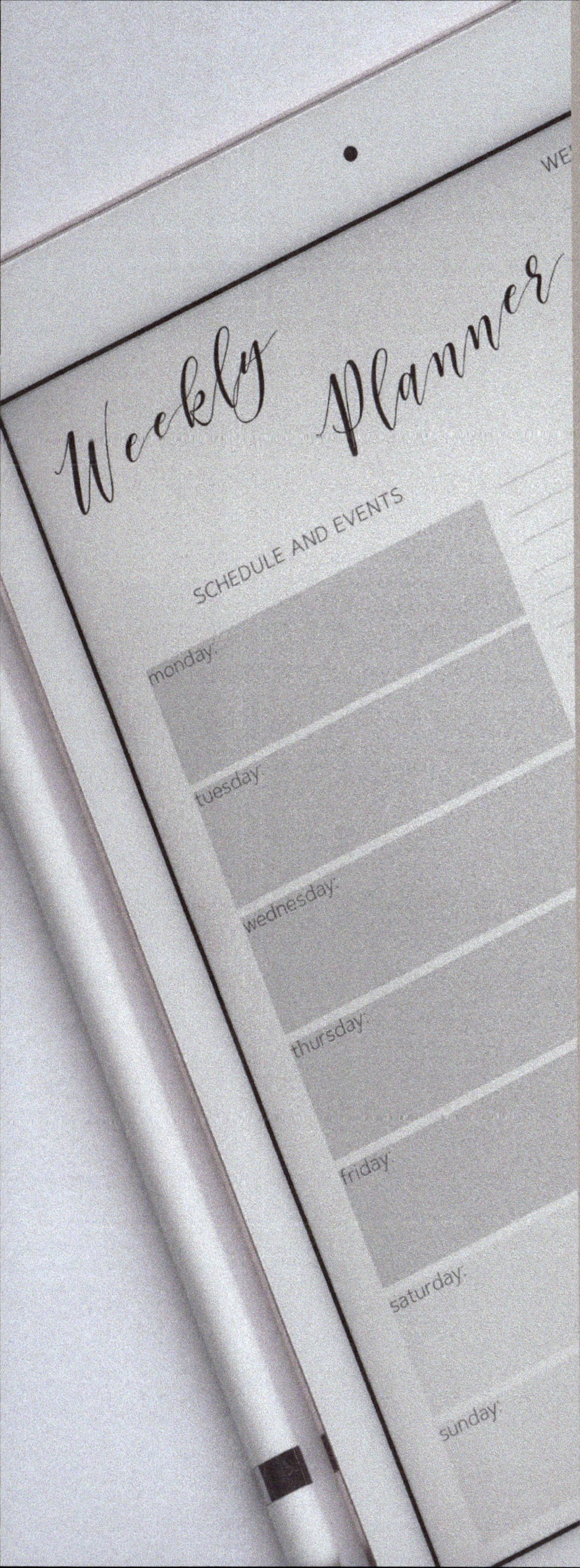

Day One

Session One

What is a matrix

Session Two

Vectors

Transposition

Session Three

Equality of matrices

Plan your work for today and every day, then work your plan

- Margaret Thatcher

SESSION ONE

WHAT IS A MATRIX

Let's get started. A matrix is a two-dimensional array of numbers, arranged in **horizontal rows** and **vertical columns.** The numbers are called elements or entries of the matrix. For example,

$$\begin{pmatrix} 1 & 2 & 3 \\ 3 & 4 & 5 \end{pmatrix} \; ; \; \begin{pmatrix} 5 \\ 6 \end{pmatrix} \; ; \; \begin{pmatrix} 0 & -7 \end{pmatrix} \; ; \; \begin{pmatrix} 1 & 7 & -5 \\ 0 & 1 & 2 \\ 3 & 7 & 15 \end{pmatrix}$$

For a beginner, a bunch of randomly arranged numbers might make no sense.

What do these arrays mean?

What are they used for?......he might think.

Surprisingly, matrices provide a powerful and elegant way to express and solve linear equations. How it does that "is one or the many mysteries that this book will reveal".

But for now, let's grapple with some matrices and understand their properties, so that the transition to linear systems become effortless.

Matrices are usually denoted using English Block letters. For example,

$$\mathrm{A} = \begin{pmatrix} 1 & 2 & 3 \\ 3 & 4 & 5 \end{pmatrix} \; ; \; \mathrm{B} = \begin{pmatrix} 5 \\ 6 \end{pmatrix} \; ; \; \mathrm{C} = \begin{pmatrix} 0 & -7 \end{pmatrix} \; ; \; \mathrm{D} = \begin{pmatrix} 1 & 7 & -5 \\ 0 & 1 & 2 \\ 3 & 7 & 15 \end{pmatrix}$$

Here the matrix A has 2 rows,

$$\mathrm{A} = \begin{pmatrix} \boxed{1 \quad 2 \quad 3} \\ \boxed{3 \quad 4 \quad 5} \end{pmatrix}$$

and 3 columns.

$$\mathrm{A} = \begin{pmatrix} \boxed{\begin{matrix} 1 \\ 3 \end{matrix}} & \boxed{\begin{matrix} 2 \\ 4 \end{matrix}} & \boxed{\begin{matrix} 3 \\ 5 \end{matrix}} \end{pmatrix}$$

Thus, A is said to have an **order** 2×3 ("2 by 3").

The matrix B has an order 2×1, since it has 2 rows,

$$\mathrm{B} = \begin{pmatrix} \boxed{5} \\ \boxed{6} \end{pmatrix}$$

and 1 column.

$$\mathrm{B} = \begin{pmatrix} \boxed{\begin{matrix} 5 \\ 6 \end{matrix}} \end{pmatrix}$$

The elements of a matrix are denoted by the subscript notation,

$$a_{ij}$$

where, *i* and *j* are the corresponding row and column to which the element belongs.

For instance, in the matrix,

$$A = \begin{pmatrix} 2 & 7 & 8 \\ -1 & 3 & 5 \\ 0 & 1 & 2 \end{pmatrix}$$

the element 2 belongs to the first row and first column (see figure 1).

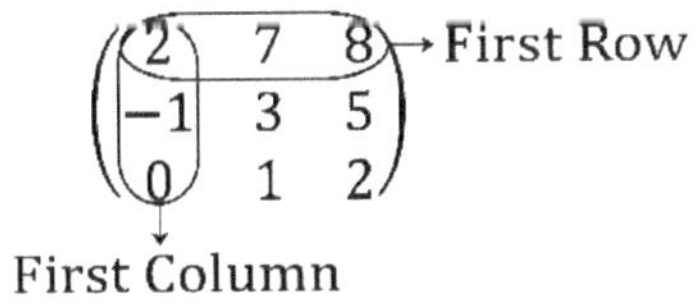

$$\begin{pmatrix} 2 & 7 & 8 \\ -1 & 3 & 5 \\ 0 & 1 & 2 \end{pmatrix}$$

First Row

First Column

Figure 1

Therefore, it is named as a_{11}

$$a_{11} = 2$$

The element 7 is named as a_{12}, since it belongs to the first row and second column (see figure 2).

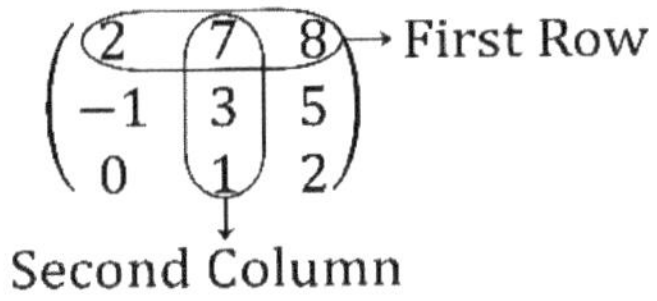

$$\begin{pmatrix} 2 & 7 & 8 \\ -1 & 3 & 5 \\ 0 & 1 & 2 \end{pmatrix}$$

First Row

Second Column

Figure 2

$$a_{12} = 7$$

Similarly, we can write,

$$a_{13} = 8 \ ; \quad a_{21} = -1 \ ; \quad a_{22} = 3 \ ; \quad a_{23} = 5 \ ; \quad a_{31} = 0 \ ; \quad a_{32} = 1 \ ; \quad a_{33} = 2$$

Thus, a general 3 × 3 matrix can be written as,

$$\begin{pmatrix} a_{11} & a_{12} & a_{13} \\ a_{21} & a_{22} & a_{23} \\ a_{31} & a_{32} & a_{33} \end{pmatrix}$$

And a general m × n matrix with m-rows and n-columns can be written as,

$$\begin{pmatrix} a_{11} & a_{12} & a_{13} & \cdots & \cdots & a_{1n} \\ a_{21} & a_{22} & a_{23} & \cdots & \cdots & a_{2n} \\ a_{31} & a_{32} & a_{33} & \cdots & \cdots & a_{3n} \\ \vdots & \vdots & \vdots & \ddots & \ddots & \vdots \\ \vdots & \vdots & \vdots & \ddots & \ddots & \vdots \\ a_{m1} & a_{m2} & a_{m3} & \cdots & \cdots & a_{mn} \end{pmatrix}$$

A matrix is **'square'**, if it has the same number of rows and columns. Example,

$$A = \begin{pmatrix} 1 & 2 & 3 \\ 4 & 5 & 6 \\ 7 & 8 & 9 \end{pmatrix} \quad ; \quad B = \begin{pmatrix} 1 & 0 \\ 0 & -1 \end{pmatrix} \quad ; \quad C = (-17)$$

order = 3 × 3 ; order = 2 × 2 ; order = 1 × 1

A general n × n square matrix with n-rows and n-columns is of the form,

$$\begin{pmatrix} a_{11} & a_{12} & \cdots & a_{1n} \\ a_{21} & a_{22} & \cdots & a_{2n} \\ \vdots & \vdots & \ddots & \vdots \\ a_{n1} & a_{n2} & \cdots & a_{nn} \end{pmatrix}$$

The diagonal containing the elements $a_{11}, a_{22}, a_{33}, \dots, a_{nn}$ is called the **'prime diagonal'** of a square matrix, and its elements are called **'diagonal elements'**. For instance, the diagonal elements of the matrix,

$$A = \begin{pmatrix} 1 & 2 & 3 \\ 4 & 5 & 6 \\ 7 & 8 & 9 \end{pmatrix}$$

are, $a_{11} = 1; a_{22} = 5; a_{33} = 9$.

A matrix is **'diagonal'**, if all the elements except the diagonal elements are zero. For example,

$$A = \begin{pmatrix} 2 & 0 \\ 0 & 3 \end{pmatrix} \quad ; \quad B = \begin{pmatrix} 15 & 0 & 0 \\ 0 & 10 & 0 \\ 0 & 0 & -2 \end{pmatrix} \quad ; \quad C = \begin{pmatrix} -7 & 0 & 0 \\ 0 & 0 & 0 \\ 0 & 0 & 5 \end{pmatrix}$$

If all the diagonal elements of a diagonal matrix are equal to unity, the matrix is called **'identity'**. All identity matrices are denoted by the letter 'I'. For example,

$$I = \begin{pmatrix} 1 & 0 & 0 \\ 0 & 1 & 0 \\ 0 & 0 & 1 \end{pmatrix} \quad ; \quad I = \begin{pmatrix} 1 & 0 \\ 0 & 1 \end{pmatrix} \quad ; \quad I = (1)$$

3 × 3 identity matrix ; 2 × 2 identity matrix ; 1 × 1 identity matrix

Problem 1: Find the orders of the following matrices.

i. $\begin{pmatrix} 1 & 2 & 3 & 4 \\ 5 & 6 & -7 & 8 \end{pmatrix}$ ii. $\begin{pmatrix} 0 & 1 \\ -1 & 0 \\ \sqrt{2} & 1 \end{pmatrix}$ iii. $\begin{pmatrix} 7 & 15 & 13 \\ 0 & 1 & 0 \\ 0 & 0 & 7 \end{pmatrix}$ iv. $(0 \quad -7)$

v. (7)

Ans:

i. 2 × 4 ii. 3 × 2 iii. 3 × 3 iv. 1 × 2 v. 1 × 1

Problem 2: If $A = \begin{pmatrix} 1 & 2 & 3 & 4 \\ 5 & 6 & 7 & 8 \\ 9 & 10 & 11 & 12 \end{pmatrix}$, find:

i. a_{23} ii. a_{33} iii. a_{31} iv. a_{14}

Ans:

i. $a_{23} = 7$ ii. $a_{33} = 11$ iii. $a_{31} = 9$ iv. $a_{14} = 4$

Problem 3: **Classify the following matrices as diagonal, identity, just square and not square (rectangular).**

i. $\begin{pmatrix} 1 & 0 \\ 0 & 3 \end{pmatrix}$ ii. $\begin{pmatrix} 1 & 1 \\ 0 & 0 \end{pmatrix}$ iii. $\begin{pmatrix} 1 & 0 & 0 \\ 0 & 1 & 0 \\ 0 & 0 & 1 \end{pmatrix}$ iv. $\begin{pmatrix} 1 & 1 & 0 \\ 0 & 1 & 0 \end{pmatrix}$

v. $\begin{pmatrix} 1 & 0 \\ 0 & -1 \end{pmatrix}$

Ans:

i. **Diagonal** ii. **Square** iii. **Identity** iv. **Rectangular**

v. **Diagonal**

Now at this point it is good to take a pause, and reflect on your progress. If you are satisfied with the topics, and was able to do the problems. Take a break, eat something if you are hungry, meditate for a while, and continue. If you are not satisfied, take a break, meditate, and find out the exact difficulty. Read think and read again until you understand it.

Remember, continue only if you are clear with the topics. Forget the one-week thing, it's a publicity stunt anyway.

SESSION TWO

Before beginning the session, remember to meditate and recall everything. Go through every topic with calm and clarity.

In this session, we first look at a special class of matrices called vectors.

VECTORS

A **vector** is a matrix that has only **one row** or **one column**. If it has **only one row**, we call it a **row vector**.

Examples of row vectors are,

$$A = \begin{pmatrix} 1 & 2 & 3 \end{pmatrix} \quad ; \quad B = \begin{pmatrix} 0 & 1 \end{pmatrix} \quad ; \quad C = \begin{pmatrix} 1 & -7 & 0 & 4 \end{pmatrix}$$

$$\text{order} = 1 \times 3 \qquad \text{order} = 1 \times 2 \qquad \text{order} = 1 \times 4$$

If the matrix has only **one column**, we call it a **column vector**. Examples of column vectors are,

$$A = \begin{pmatrix} 1 \\ -1 \\ 2 \end{pmatrix} \quad ; \quad B = \begin{pmatrix} 1 \\ 2 \end{pmatrix} \quad ; \quad C = \begin{pmatrix} 0 \\ -7 \\ 2 \\ 4 \end{pmatrix}$$

$$\text{order} = 3 \times 1 \qquad \text{order} = 2 \times 1 \qquad \text{order} = 4 \times 1$$

TRANSPOSITION

A column vector can be converted into a row vector and vice versa using an operation called **'transposition'**.

$$\begin{pmatrix} \text{column} \\ \text{vector} \end{pmatrix} \xrightarrow{\text{transposition}} (\text{row vector})$$

$$\text{order} = \mathrm{n} \times 1 \qquad \text{order} = 1 \times \mathrm{n}$$

$$(\text{row vector}) \xrightarrow{\text{transposition}} \begin{pmatrix} \text{column} \\ \text{vector} \end{pmatrix}$$

$$\text{order} = 1 \times \mathrm{n} \qquad \text{order} = \mathrm{n} \times 1$$

For example, if A is the column vector,

$$A = \begin{pmatrix} 1 \\ 2 \\ 3 \end{pmatrix}$$

$$\text{order} = 3 \times 1$$

Then A^T ("A-transpose") is the row vector obtained by converting the column into a row. i.e.,

$$A^T = \begin{pmatrix} 1 & 2 & 3 \end{pmatrix}$$

$$\text{order} = 1 \times 3$$

$$A = \begin{pmatrix} 1 \\ 2 \\ 3 \end{pmatrix} \xrightarrow{\text{transposition}} \begin{pmatrix} 1 & 2 & 3 \end{pmatrix} = A^T$$

$$\text{order} = 1 \times 3$$

$$\text{order} = 3 \times 1$$

and if 'A' is a row vector. For example,

$$A = \begin{pmatrix} 0 & 2 & -3 \end{pmatrix}$$

$$\text{order} = 1 \times 3$$

Then A^T is the column vector obtained by converting the row into a column. i.e.,

$$A^T = \begin{pmatrix} 0 \\ 2 \\ -3 \end{pmatrix}$$

$$\text{order} = 3 \times 1$$

$$A = \begin{pmatrix} 0 & 2 & -3 \end{pmatrix} \xrightarrow{\text{transposition}} \begin{pmatrix} 0 \\ 2 \\ -3 \end{pmatrix} = A^T$$

$$\text{order} = 1 \times 3$$

$$\text{order} = 3 \times 1$$

To summarize,

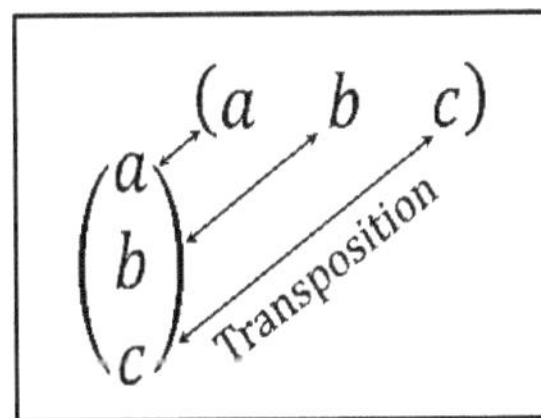

The definition of transposition is not restricted to vectors alone. In fact, any matrix can be transposed by **converting its columns into rows one by one**. For example, if,

then,

$$A = \begin{pmatrix} 1 & 4 & 7 \\ 2 & 5 & 8 \\ 3 & 6 & 9 \end{pmatrix} \qquad A^T = \begin{pmatrix} 1 & 2 & 3 \\ 4 & 5 & 6 \\ 7 & 8 & 9 \end{pmatrix}$$

$$\text{order} = 3 \times 3 \qquad \text{order} = 3 \times 3$$

or if,

then,

$$B = \begin{pmatrix} 1 & 0 \\ 4 & 8 \\ 2 & 9 \end{pmatrix} \qquad B^T = \begin{pmatrix} 1 & 4 & 2 \\ 0 & 8 & 9 \end{pmatrix}$$

order = 3 × 2 order = 2 × 3

A matrix is **'symmetric'** if it is equal to its transpose. That is,

$$\boxed{A^T = A}$$

Examples are,

$$A = \begin{pmatrix} 1 & 8 \\ 8 & -7 \end{pmatrix} \quad ; \quad A^T = \begin{pmatrix} 1 & 8 \\ 8 & -7 \end{pmatrix}$$

$$B = \begin{pmatrix} 0 & 8 & 3 \\ 8 & 1 & 5 \\ 3 & 5 & 2 \end{pmatrix} \quad ; \quad B^T = \begin{pmatrix} 0 & 8 & 3 \\ 8 & 1 & 5 \\ 3 & 5 & 2 \end{pmatrix}$$

Note that the transpose of a transpose is the matrix itself. That is,

$$\boxed{(A^T)^T = A}$$

Problem 4: Find the transposes of the following matrices.

i. $\begin{pmatrix} 7 \\ 15 \\ 1 \\ 9 \end{pmatrix}$ ii. $\begin{pmatrix} 1 & 2 \\ 7 & 8 \\ 14 & 0 \end{pmatrix}$ iii. $\begin{pmatrix} 1 & 0 & 0 \\ 0 & 1 & 0 \\ 0 & 0 & 1 \end{pmatrix}$ iv. $\begin{pmatrix} 0 & 15 & 8 \end{pmatrix}$

v. $\begin{pmatrix} 1 & 0 \\ -1 & 0 \end{pmatrix}$

Ans:

i. $\begin{pmatrix} 7 & 15 & 1 & 9 \end{pmatrix}$ ii. $\begin{pmatrix} 1 & 7 & 14 \\ 2 & 8 & 0 \end{pmatrix}$ iii. $\begin{pmatrix} 1 & 0 & 0 \\ 0 & 1 & 0 \\ 0 & 0 & 1 \end{pmatrix}$

iv. $\begin{pmatrix} 0 \\ 15 \\ 8 \end{pmatrix}$ v. $\begin{pmatrix} 1 & -1 \\ 0 & 0 \end{pmatrix}$

SESSION THREE

EQUALITY OF MATRICES

In the previous session, we have seen that a symmetric matrix is equal to its transpose. But since matrices are not numbers, how can we define their equality.

Here's how,

Two matrices of the same order are said to be equal, if their corresponding elements are equal.

For instance, if the matrices,

$$A = \begin{pmatrix} a_{11} & a_{12} \\ a_{21} & a_{22} \end{pmatrix} \text{ and } B = \begin{pmatrix} b_{11} & b_{12} \\ b_{21} & b_{22} \end{pmatrix}$$

are equal, then,

$$\begin{matrix} a_{11} = b_{11} & a_{12} = b_{12} \\ a_{21} = b_{21} & a_{22} = b_{22} \end{matrix}$$

Example 1: If

$$\begin{pmatrix} x+y & 3 \\ 2 & 9 \end{pmatrix} = \begin{pmatrix} 5 & 3 \\ y & 9 \end{pmatrix},$$

find x and y.

Solution: Since the two matrices are equal, we get the equations†,

$$\left.\begin{matrix} x + y = 5 \\ y = 2 \end{matrix}\right\}$$

Upon solving, we get,

$$\left.\begin{matrix} x = 3 \\ y = 2 \end{matrix}\right\}$$

† If you don't know how to solve linear equations, see Day-IV

Problem 5: If $\begin{pmatrix} 1 & 7 & 8 \\ b & 1 & 2 \\ 8 & -1 & a+b+c \end{pmatrix} = \begin{pmatrix} 1 & 7 & 8 \\ 0 & 1 & a+b \\ 8 & -1 & 4 \end{pmatrix}$ **Find a, b and c.**

Ans:

$a = 2, \; b = 0, \; c = 2$

Problem 6: If

$$A = \begin{pmatrix} 1 & 2 & x \\ y & 5 & 9 \\ 1 & 9 & 10 \end{pmatrix}$$

is symmetric, find x and y.

Ans:

$x = 1,\ y = 2$

Before ending the session, remember to meditate and recall everything. If you are satisfied with the sessions, you are good to go. Have a good night sleep and comeback tomorrow. If you are not satisfied, it's ok. Tomorrow, return to the same session.

Don't give up.

Day Two

Session One

Matrix addition

Multiplication by a number

Session Two

Product of two matrices

Session Three

Matrix multiplication continued

I hate every minute of my training.
But I said, don't quit. Suffer now and
live the rest of your life as a champion.

- Muhammad Ali

SESSION ONE

Take a few breaths and meditate on everything we have studied so far. If you have flash cards, go through them with patience and focus. Try to recall and reconnect with each and every topic.

Today, we will learn the basic operations involving matrices. They are:

i. Addition of two matrices
ii. Multiplication of a matrix by a number (Scalar Multiplication)
iii. Multiplication of a matrix by another matrix

MATRIX ADDITION

Addition of two matrices A and B are conformable, if and only if they are of the same order. Their sum A + B is the matrix obtained by adding the corresponding elements of A and B, and has the same order of that of A and B. That is, if

$$A = \begin{pmatrix} a_{11} & a_{12} & \cdots & \cdots & a_{1n} \\ a_{21} & a_{22} & \cdots & \cdots & a_{2n} \\ \vdots & \vdots & \ddots & \ddots & \vdots \\ \vdots & \vdots & \ddots & \ddots & \vdots \\ a_{m1} & a_{m2} & \cdots & \cdots & a_{mn} \end{pmatrix} \quad \text{and} \quad B = \begin{pmatrix} b_{11} & b_{12} & \cdots & \cdots & b_{1n} \\ b_{21} & b_{22} & \cdots & \cdots & b_{2n} \\ \vdots & \vdots & \ddots & \ddots & \vdots \\ \vdots & \vdots & \ddots & \ddots & \vdots \\ b_{m1} & b_{m2} & \cdots & \cdots & b_{mn} \end{pmatrix}$$

$$\text{order} = \text{m} \times \text{n} \qquad\qquad \text{order} = \text{m} \times \text{n}$$

then,

$$A + B = \begin{pmatrix} a_{11} + b_{11} & a_{12} + b_{12} & \cdots & \cdots & a_{1n} + b_{1n} \\ a_{21} + b_{21} & a_{22} + b_{22} & \cdots & \cdots & a_{2n} + b_{2n} \\ \vdots & \vdots & \ddots & \ddots & \vdots \\ \vdots & \vdots & \ddots & \ddots & \vdots \\ a_{m1} + b_{m1} & a_{m2} + b_{m2} & \cdots & \cdots & a_{mn} + b_{mn} \end{pmatrix}$$

$$\text{order} = \text{m} \times \text{n}$$

For example, if,

$$A = \begin{pmatrix} 1 & 2 \\ 7 & -1 \\ 2 & 3 \end{pmatrix} \quad \text{and} \quad B = \begin{pmatrix} 1 & 2 \\ 4 & 8 \\ 3 & 1 \end{pmatrix}$$

$$\text{order} = 3 \times 2 \qquad\qquad \text{order} = 3 \times 2$$

then,

$$A + B = \begin{pmatrix} 1+1 & 2+2 \\ 7+4 & -1+8 \\ 2+3 & 3+1 \end{pmatrix} = \begin{pmatrix} 2 & 4 \\ 11 & 7 \\ 5 & 4 \end{pmatrix}$$

$$\text{order} = 3 \times 2 \qquad \text{order} = 3 \times 2$$

Addition of the matrices,

$$A = \begin{pmatrix} 2 & -2 \\ 1 & 3 \end{pmatrix} \text{ and } B = \begin{pmatrix} 3 & 7 & -1 \\ 2 & 15 & 2 \end{pmatrix}$$

is not conformable, since their orders are different.

Also, similar to matrix addition, the **difference of two matrices is obtained by taking the difference of their corresponding elements**. For example, if,

$$A = \begin{pmatrix} 3 & 7 \\ 1 & 8 \end{pmatrix} \text{ and } B = \begin{pmatrix} 7 & 1 \\ 2 & -4 \end{pmatrix}$$

then,

$$A - B = \begin{pmatrix} 3-7 & 7-1 \\ 1-2 & 8-(-4) \end{pmatrix} = \begin{pmatrix} -4 & 6 \\ -1 & 12 \end{pmatrix}$$

and,

$$B - A = \begin{pmatrix} 7-3 & 1-7 \\ 2-1 & -4-8 \end{pmatrix} = \begin{pmatrix} 4 & -6 \\ 1 & -12 \end{pmatrix}$$

Problem 7: If

$$A = \begin{pmatrix} 7 & -13 & 1 \\ 0 & 5 & 4 \\ 3 & -2 & 1 \end{pmatrix} \text{ and } B = \begin{pmatrix} 15 & -14 & 0 \\ -1 & 0 & 1 \\ 1 & 7 & 4 \end{pmatrix}$$

find:

i. $A + B$ **ii.** $A - B$ **iii.** $B - A$

Ans:

i. $\begin{pmatrix} 22 & -27 & 1 \\ -1 & 5 & 5 \\ 4 & 5 & 5 \end{pmatrix}$ **ii.** $\begin{pmatrix} -8 & 1 & 1 \\ 1 & 5 & 3 \\ 2 & -9 & -3 \end{pmatrix}$ **iii.** $\begin{pmatrix} 8 & -1 & -1 \\ -1 & -5 & -3 \\ -2 & 9 & 3 \end{pmatrix}$

Problem 8: If

$$\mathbf{A} = \begin{pmatrix} 3 & 7 \\ 1 & 8 \end{pmatrix} \text{ and } \mathbf{B} = \begin{pmatrix} 7 & 1 \\ 2 & -4 \end{pmatrix}$$

show that matrix addition is commutative. That is, show that:

$$\mathbf{A + B = B + A}$$

MULTIPLICATION BY A NUMBER

The product of a matrix 'A' by a number 'k' is obtained by multiplying every element of A by the number k. For example, if

$$A = \begin{pmatrix} 1 & 2 & 3 \\ 7 & -4 & 2 \end{pmatrix} \text{ and } k = -2$$

then,

$$kA = -2\begin{pmatrix} 1 & 2 & 3 \\ 7 & -4 & 2 \end{pmatrix} = \begin{pmatrix} -2(1) & -2(2) & -2(3) \\ -2(7) & -2(-4) & -2(2) \end{pmatrix} = \begin{pmatrix} -2 & -4 & -6 \\ -14 & 8 & -4 \end{pmatrix}$$

Conversely, a number can be factored out of a matrix. For example,

$$\begin{pmatrix} 2 & -4 & 8 \\ -4 & 6 & 10 \\ 2 & 3 & -12 \end{pmatrix} = -2\begin{pmatrix} -1 & 2 & -4 \\ 2 & -3 & -5 \\ -1 & -3/2 & 6 \end{pmatrix}$$

A matrix is **'anti-symmetric'**, if its transpose is equal to its negative.

i.e.

$$\boxed{A^T = -A}$$

For example, if

$$A = \begin{pmatrix} 0 & -1 & 3 \\ 1 & 0 & -2 \\ -3 & 2 & 0 \end{pmatrix}$$

then,

$$A^T = \begin{pmatrix} 0 & 1 & -3 \\ -1 & 0 & 2 \\ 3 & -2 & 0 \end{pmatrix}$$

Factoring '−1' out of the matrix gives,

$$A^T = -\begin{pmatrix} 0 & -1 & 3 \\ 1 & 0 & -2 \\ -3 & 2 & 0 \end{pmatrix} = -A$$

i.e.

$$A^T = -A$$

Therefore, A is anti-symmetric.

Problem 9: If $A = \begin{pmatrix} 7 & 1 \\ 0 & 15 \end{pmatrix}$ and $B = \begin{pmatrix} 2 & -4 \\ 5 & 9 \end{pmatrix}$, find:

i. 2A ii. 3B iii. 2A + 3B iv. 2A − 3B

Ans:

i. $\begin{pmatrix} 14 & 2 \\ 0 & 30 \end{pmatrix}$ ii. $\begin{pmatrix} 6 & -12 \\ 15 & 27 \end{pmatrix}$ iii. $\begin{pmatrix} 20 & -10 \\ 15 & 57 \end{pmatrix}$ iv. $\begin{pmatrix} 8 & 14 \\ -15 & 3 \end{pmatrix}$

Problem 10: If

$$A = \begin{pmatrix} 1 & 2 & -4 \\ 3 & 8 & 5 \\ -2 & 4 & 7 \end{pmatrix}$$

show that $A + A^T$ is symmetric and $A - A^T$ is anti-symmetric.

Operations involving matrices are more prone to errors than those with numbers. So, execute them carefully, and double check the answer every time.

But for now, take a few minutes off, refresh yourself, meditate and then comeback.

SESSION TWO

PRODUCT OF TWO MATRICES

This session is divided into two parts.

The first part is concerned with the conformability of matrix multiplication. Where we detail the conditions for a product to exist. In the second part, we present the multiplication itself.

PART ONE – CONFORMABILITY

The product of two matrices exists if and only if, the number of columns of the first matrix is equal to the number of rows of the second.

If A has an order $m \times k$ and B has an order $k \times n$, then their product is conformable, since the number of columns of A and the number of rows of B are equal.

$$\underset{m \times \boxed{k = k} \times n}{(A)\,(B)}$$

no. of columns of A = no. of rows of B

Their product (AB) has an order equal to the number of rows of A (i.e., m) and the number of columns of B (i.e., n).

$$\underset{m \times \boxed{k = k} \times n}{(A)\,(B)} = \underset{m \times n}{(AB)}$$

For example, if A has an order 2×3 and B has an order 3×1. Then their product (AB) is conformable and has an order 2×1. That is,

$$\underset{2 \times \boxed{3 = 3} \times 1}{(A)\,(B)} = \underset{2 \times 1}{(AB)}$$

But if, A has an order 3×2 and B has an order 1×3, their product (AB) is not conformable, since the number of columns of A is not equal to the number of rows of B. i.e.,

$$\underset{3 \times \boxed{2 \neq 1} \times 3}{(A)\,(B)}$$

But the product (BA) is possible, with an order 1×2.

$$\underset{1 \times \boxed{3 = 3} \times 2}{(B)\,(A)} = \underset{1 \times 2}{(BA)}$$

Now, if A has an order 2×3 and B has an order 3×2. Their products (AB) and (BA) will have two different orders.

$$\underset{2 \times \boxed{3 = 3} \times 2}{(A)\,(B)} = \underset{2 \times 2}{(AB)}$$

and,

$$\underset{3 \times \boxed{2 = 2} \times 3}{(B)\,(A)} = \underset{3 \times 3}{(BA)}$$

This shows that matrix multiplication is not necessarily commutative.

Problem 11: If $A = \begin{pmatrix} 1 & 2 \\ 3 & 1 \end{pmatrix}$; $B = \begin{pmatrix} 1 \\ 2 \\ 3 \end{pmatrix}$; $C = (7 \quad 1 \quad -1)$;

$D = \begin{pmatrix} 1 & 7 & 8 \\ 5 & 9 & 0 \\ 0 & -1 & 15 \end{pmatrix}$; $E = \begin{pmatrix} 0 \\ 1 \end{pmatrix}$ and $F = \begin{pmatrix} -7 & 5 & 1 \\ 0 & 1 & 18 \\ 8 & 10 & 1 \end{pmatrix}$.

Find the orders of the following products.

i.	**CB**	**ii.**	**CD**	**iii.**	**DF**	**iv.**	**AB**
v.	**CE**	**vi.**	**AE**	**vii.**	**EA**	**viii.**	**BC**

Ans:

i.	**1×1**	**ii.**	**1×3**
iii.	**3×3**	**iv.**	**Not Conformable**
v.	**Not Conformable**	**vi.**	**2×1**
vii.	**Not Conformable**	**viii.**	**3×3**

PART TWO – MATRIX MULTIPLICATION

Matrix multiplication might be the most unnatural operation you might ever encounter in this entire course.

Let's begin with a simple case.

Consider the multiplication of the row and the column vectors,

$$A = (a_{11} \quad a_{12}) \quad \text{and} \quad B = \begin{pmatrix} b_{11} \\ b_{21} \end{pmatrix}$$

$$1 \times 2 \qquad\qquad 2 \times 1$$

with the row vector to the left of the column vector. i.e.,

$$(A)\,(B) = (a_{11} \quad a_{12}) \begin{pmatrix} b_{11} \\ b_{21} \end{pmatrix}$$

$$1 \times \boxed{2 = 2} \times 1 \qquad\qquad 1 \times \boxed{2 = 2} \times 1$$

Their product AB should have an order 1×1.

That is, it should contain only one element.

$$\underset{1\times\boxed{2=2}\times 1}{(A)\,(B)} = \underset{1\times\boxed{2\;=\;2}\times 1}{\begin{pmatrix} a_{11} & a_{12} \end{pmatrix}\begin{pmatrix} b_{11} \\ b_{21} \end{pmatrix}} = \underset{1\times 1}{(\,?\,)}$$

Now comes the fun part,

If you are a right-handed writer, place your index finger of your left hand on a_{11} and your pen on b_{11} (see figure 3).

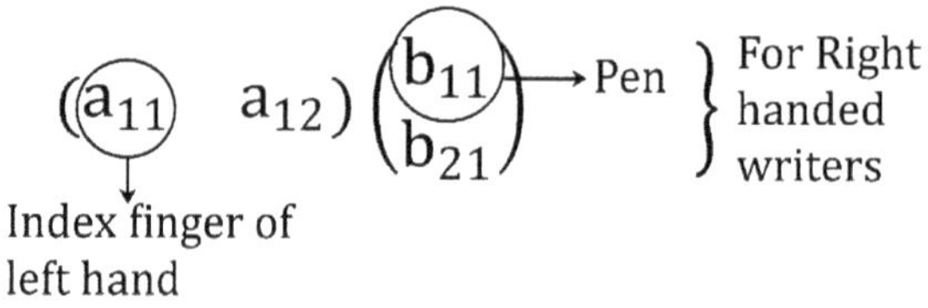

$$\begin{pmatrix} \textcircled{a_{11}} & a_{12} \end{pmatrix}\begin{pmatrix} \textcircled{b_{11}} \\ b_{21} \end{pmatrix}$$

a_{11} → Index finger of left hand; b_{11} → Pen } For Right handed writers

Figure 3

If you are a left-handed writer, place your pen on a_{11} and your index finger of your right hand on b_{11}(see figure 4).

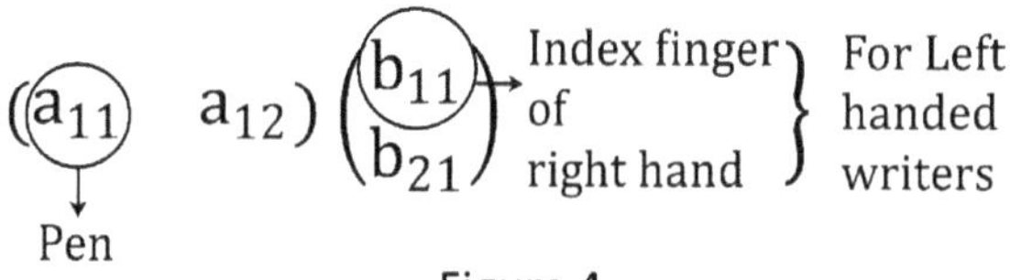

$$\begin{pmatrix} \textcircled{a_{11}} & a_{12} \end{pmatrix}\begin{pmatrix} \textcircled{b_{11}} \\ b_{21} \end{pmatrix}$$

a_{11} → Pen; b_{11} → Index finger of right hand } For Left handed writers

Figure 4

Multiply those two elements to obtain

$$a_{11}b_{11}$$

Now, move the positions of your finger and your pen to a_{12} and b_{21} (see figure 5).

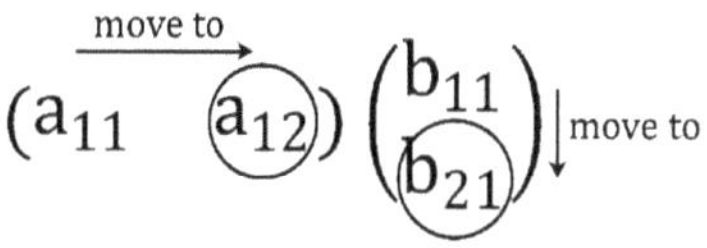

$$\begin{pmatrix} a_{11} & \textcircled{a_{12}} \end{pmatrix}\begin{pmatrix} b_{11} \\ \textcircled{b_{21}} \end{pmatrix}$$

a_{11} move to a_{12}; b_{11} move to b_{21}

Figure 5

and multiply those together to obtain

$$a_{12}b_{21}$$

The sum of those two terms gives the element in the product matrix. That is,

$$\underset{1\times\boxed{2\;=\;2}\times 1}{\begin{pmatrix} a_{11} & a_{12} \end{pmatrix}\begin{pmatrix} b_{11} \\ b_{21} \end{pmatrix}} = \underset{1\times 1}{\begin{pmatrix} a_{11}b_{11} + a_{12}b_{21} \end{pmatrix}}$$

Let us call the product matrix C, and its element c_{11}. That is,

$$AB = C$$

or,

$$\begin{pmatrix} a_{11} & a_{12} \end{pmatrix} \begin{pmatrix} b_{11} \\ b_{21} \end{pmatrix} = (c_{11})$$

where,

$$c_{11} = a_{11}b_{11} + a_{12}b_{21}$$

To summarize,

The product of a row and a column vector with the row vector to the left of the column vector is obtained by multiplying the elements at each position and adding them.

For example, if

$$A = \begin{pmatrix} 1 & 2 \end{pmatrix} \text{ and } B = \begin{pmatrix} 3 \\ 4 \end{pmatrix}$$

Then,

$$AB = \begin{pmatrix} 1 & 2 \end{pmatrix} \begin{pmatrix} 3 \\ 4 \end{pmatrix} = \big((1)(3) + (2)(4)\big) = (11)$$

This method can be extended to row and column vectors of all orders. That is,

$$\begin{pmatrix} a_{11} & a_{12} & \cdots & a_{1k} \end{pmatrix} \begin{pmatrix} b_{11} \\ b_{21} \\ \vdots \\ b_{k1} \end{pmatrix} = \underset{1 \times 1}{(a_{11}b_{11} + a_{12}b_{21} + \cdots + a_{1k}b_{k1})}$$

$$1 \times \boxed{k \quad = \quad k} \times 1$$

For example, if

$$A = \begin{pmatrix} 1 & 2 & 3 & 4 \end{pmatrix} \text{ and } B = \begin{pmatrix} 5 \\ 6 \\ 7 \\ 8 \end{pmatrix}$$

then,

$$AB = \begin{pmatrix} 1 & 2 & 3 & 4 \end{pmatrix} \begin{pmatrix} 5 \\ 6 \\ 7 \\ 8 \end{pmatrix} = \big((1)(5) + (2)(6) + (3)(7) + (4)(8)\big) = (70)$$

Problem 12: If $A = \begin{pmatrix} 1 & -1 & 0 \end{pmatrix}$ and $B = \begin{pmatrix} 7 \\ 8 \\ 9 \end{pmatrix}$, Find:

i. $2A$ ii. $2A + B$ iii. $(2A)(3B)$

Ans:

i. $\begin{pmatrix} 2 & -2 & 0 \end{pmatrix}$ ii. Addition Not Conformable iii. (-6)

Now, let's add another column to the matrix B,

$$\begin{pmatrix} a_{11} & a_{12} & \cdots & a_{1k} \end{pmatrix} \begin{pmatrix} b_{11} & b_{12} \\ b_{21} & b_{22} \\ \vdots & \vdots \\ b_{k1} & b_{k2} \end{pmatrix} = \underset{1 \times 2}{\begin{pmatrix} ? & ? \end{pmatrix}}$$

$$1 \times \boxed{k \quad = \quad k} \times 2$$

Here, the product should contain two elements since it has an order 1×2.

The first one is in the first row-first column (c_{11}) and the second one is in the first row-second column (c_{12}). That is,

$$\begin{pmatrix} a_{11} & a_{12} & \cdots & a_{1k} \end{pmatrix} \begin{pmatrix} b_{11} & b_{12} \\ b_{21} & b_{22} \\ \vdots & \vdots \\ b_{k1} & b_{k2} \end{pmatrix} = \begin{pmatrix} c_{11} & c_{12} \end{pmatrix}$$

Similar to the previous case, c_{11} is calculated by multiplying and adding the corresponding elements of the first row and first column of A and B respectively. That is,

$$\begin{pmatrix} \boxed{a_{11} \quad a_{12} \quad \cdots \quad a_{1k}} \end{pmatrix} \begin{pmatrix} \boxed{\begin{matrix} b_{11} \\ b_{21} \\ \vdots \\ b_{k1} \end{matrix}} & \begin{matrix} b_{12} \\ b_{22} \\ \vdots \\ b_{k2} \end{matrix} \end{pmatrix} = \begin{pmatrix} \boxed{c_{11}} & c_{12} \end{pmatrix}$$

$$c_{11} = a_{11}b_{11} + a_{12}b_{21} + \cdots + a_{1k}b_{k1}$$

where 'k' is any positive integer.

Now the second element c_{12} is calculated by multiplying and adding the corresponding elements of the first row of A and the second column of B. Hence the name-c_{12}.

$$c_{1\,2}$$

1st row of A   2nd column of B

That is,

$$\left(\boxed{a_{11} \quad a_{12} \quad \cdots \quad a_{1k}}\right)\begin{pmatrix} b_{11} & \boxed{\begin{matrix} b_{12} \\ b_{22} \\ \vdots \\ b_{k2} \end{matrix}} \\ b_{21} & \\ \vdots & \\ b_{k1} & \end{pmatrix} = \left(c_{11} \quad \boxed{c_{12}}\right)$$

$$c_{12} = a_{11}b_{12} + a_{12}b_{22} + \cdots + a_{1k}b_{k2}$$

For example,

$$(1 \quad 2)\begin{pmatrix} 3 & 5 \\ 4 & 6 \end{pmatrix} = ((1)(3) + (2)(4) \quad (1)(5) + (2)(6)) = (11 \quad 17)$$

This method can be extended for any number of columns of B. i.e.,

$$(a_{11} \quad a_{12} \quad \cdots \quad a_{1k})\begin{pmatrix} b_{11} & b_{12} & \cdots & b_{1n} \\ b_{21} & b_{22} & \cdots & b_{2n} \\ \vdots & \vdots & \ddots & \vdots \\ b_{k1} & b_{k2} & \cdots & b_{kn} \end{pmatrix} = \underset{1\times n}{(c_{11} \quad c_{12} \quad \cdots \quad c_{1n})}$$

$$1 \times \boxed{k \qquad = \qquad k} \times n$$

For example,

$$(3 \quad -1 \quad 7)\begin{pmatrix} 1 & 2 & 7 & -2 \\ 0 & 8 & -2 & 3 \\ -1 & 3 & 9 & 5 \end{pmatrix} = (-4 \quad 19 \quad 86 \quad 26)$$

Problem 13: If $A = (1 \quad -2 \quad 3 \quad 4)$; $B = \begin{pmatrix} -1 & 2 & 3 \\ 1 & 7 & 0 \\ 2 & 15 & 0 \end{pmatrix}$; $C = \begin{pmatrix} -1 & 0 & -1 & 1 \\ 0 & 1 & -1 & 0 \\ 0 & 1 & 3 & 5 \\ 1 & 2 & 4 & 6 \end{pmatrix}$

and $D = (7 \quad 8 \quad 9)$. **Find:**

i. (2A)(C) **ii.** AD **iii.** (D)(−B) **iv.** DC

Ans:

i. $(6 \quad 18 \quad 52 \quad 80)$ **ii.** Not Conformable

iii. $(-19 \quad -205 \quad -21)$ **iv.** Not Conformable

Let's take a break and refresh ourselves.

Take some time to meditate. The next session is an extension of this one. So, continue only if you are clear with the topics.

SESSION THREE

MATRIX MULTIPLICATION- CONTINUED

In the previous session, the matrix A in the product AB contained only one row. In this session, let's add another.

$$\begin{pmatrix} a_{11} & a_{12} & \cdots & a_{1k} \\ a_{21} & a_{22} & \cdots & a_{2k} \end{pmatrix} \begin{pmatrix} b_{11} & b_{12} & \cdots & b_{1n} \\ b_{21} & b_{22} & \cdots & b_{2n} \\ \vdots & \vdots & \ddots & \vdots \\ b_{k1} & b_{k2} & \cdots & b_{kn} \end{pmatrix} = \begin{pmatrix} c_{11} & c_{12} & \cdots & c_{1n} \\ c_{21} & c_{22} & \cdots & c_{2n} \end{pmatrix}$$

$$2 \times \boxed{k \quad = \quad k} \times n \qquad\qquad 2 \times n$$

Here, the product contains two rows, since it has an order $2 \times n$. This is due to the presence of an additional row in the matrix A.

The first row in the product is calculated using the method detailed in the previous session.

i.e., multiplying and adding the elements of the first row of A with each column of B.

$$\begin{pmatrix} \boxed{a_{11} \quad a_{12} \quad \cdots \quad a_{1k}} \\ a_{21} \quad a_{22} \quad \cdots \quad a_{2k} \end{pmatrix} \begin{pmatrix} \boxed{\begin{matrix} b_{11} \\ b_{21} \\ \vdots \\ b_{k1} \end{matrix}} & \boxed{\begin{matrix} b_{12} \\ b_{22} \\ \vdots \\ b_{k2} \end{matrix}} & \begin{matrix} \cdots \\ \cdots \\ \ddots \\ \cdots \end{matrix} & \boxed{\begin{matrix} b_{1n} \\ b_{2n} \\ \vdots \\ b_{kn} \end{matrix}} \end{pmatrix} = \begin{pmatrix} \boxed{c_{11} \quad c_{12} \quad \cdots \quad c_{1n}} \\ c_{21} \quad c_{22} \quad \cdots \quad c_{2n} \end{pmatrix}$$

Now the second row in the product is calculated by multiplying and adding the corresponding elements of the second row of A with each column of B. i.e.,

$$\begin{pmatrix} a_{11} \quad a_{12} \quad \cdots \quad a_{1k} \\ \boxed{a_{21} \quad a_{22} \quad \cdots \quad a_{2k}} \end{pmatrix} \begin{pmatrix} \boxed{\begin{matrix} b_{11} \\ b_{21} \\ \vdots \\ b_{k1} \end{matrix}} & \boxed{\begin{matrix} b_{12} \\ b_{22} \\ \vdots \\ b_{k2} \end{matrix}} & \begin{matrix} \cdots \\ \cdots \\ \ddots \\ \cdots \end{matrix} & \boxed{\begin{matrix} b_{1n} \\ b_{2n} \\ \vdots \\ b_{kn} \end{matrix}} \end{pmatrix} = \begin{pmatrix} c_{11} \quad c_{12} \quad \cdots \quad c_{1n} \\ \boxed{c_{21} \quad c_{22} \quad \cdots \quad c_{2n}} \end{pmatrix}$$

For example,

$$\begin{pmatrix} 1 & 7 \\ 0 & 2 \end{pmatrix}\begin{pmatrix} 7 & 8 \\ -1 & 5 \end{pmatrix} = \begin{pmatrix} (1)(7)+(7)(-1) & (1)(8)+(7)(5) \\ (0)(7)+(2)(-1) & (0)(8)+(2)(5) \end{pmatrix} = \begin{pmatrix} 0 & 43 \\ -2 & 10 \end{pmatrix}$$

$$2 \times \boxed{2 = 2} \times 2 \qquad\qquad 2 \times 2 \qquad\qquad 2 \times 2$$

This method can be extended for any number of rows of A. i.e.,

$$\begin{pmatrix} a_{11} & a_{12} & \cdots & a_{1k} \\ a_{21} & a_{22} & \cdots & a_{2k} \\ \vdots & \vdots & \ddots & \vdots \\ a_{m1} & a_{m2} & \cdots & a_{mk} \end{pmatrix}\begin{pmatrix} b_{11} & b_{12} & \cdots & b_{1n} \\ b_{21} & b_{22} & \cdots & b_{2n} \\ \vdots & \vdots & \ddots & \vdots \\ b_{k1} & b_{k2} & \cdots & b_{kn} \end{pmatrix} = \begin{pmatrix} c_{11} & c_{12} & \cdots & c_{1n} \\ c_{21} & c_{22} & \cdots & c_{2n} \\ \vdots & \vdots & \ddots & \vdots \\ c_{m1} & c_{m2} & \cdots & c_{mn} \end{pmatrix}$$

$$m \times \boxed{k \qquad = \qquad k} \times n \qquad\qquad m \times n$$

For example,

$$\begin{pmatrix} 1 & 7 & -5 \\ 0 & 1 & 0 \\ 2 & 4 & 3 \end{pmatrix}\begin{pmatrix} 1 & 0 \\ 2 & -1 \\ 4 & 5 \end{pmatrix} = \begin{pmatrix} (1)(1)+(7)(2)+(-5)(4) & (1)(0)+(7)(-1)+(-5)(5) \\ (0)(1)+(1)(2)+(0)(4) & (0)(0)+(1)(-1)+(0)(5) \\ (2)(1)+(4)(2)+(3)(4) & (2)(0)+(4)(-1)+(3)(5) \end{pmatrix}$$

$$3 \times \boxed{3 \; = \; 3} \times 2 \qquad\qquad 3 \times 2$$

$$= \begin{pmatrix} -5 & -32 \\ 2 & -1 \\ 22 & 11 \end{pmatrix}$$

$$3 \times 2$$

Problem 14: If $\mathbf{A = \begin{pmatrix} 1 & -2 \\ 3 & -4 \end{pmatrix}; B = \begin{pmatrix} 1 & 8 & 7 \\ 5 & 2 & 4 \\ -1 & 0 & 0 \end{pmatrix}; C = \begin{pmatrix} 1 & 1 & 4 \\ 0 & 1 & 2 \\ 8 & 0 & -1 \end{pmatrix};}$

$\mathbf{D = \begin{pmatrix} 3 & 4 & 5 \\ 2 & -1 & 8 \end{pmatrix}}$ **and** $\mathbf{E = \begin{pmatrix} 7 & 8 \\ 9 & 10 \end{pmatrix}}$**, find:**

i. AD ii. EA iii. (E)(A^T) iv. BC v. ED vi. DE

Ans:

i. $\begin{pmatrix} -1 & 6 & -11 \\ 1 & 16 & -17 \end{pmatrix}$ **ii.** $\begin{pmatrix} 31 & -46 \\ 39 & -58 \end{pmatrix}$ **iii.** $\begin{pmatrix} -9 & -11 \\ -11 & -13 \end{pmatrix}$

iv. $\begin{pmatrix} 57 & 9 & 13 \\ 37 & 7 & 20 \\ -1 & -1 & -4 \end{pmatrix}$ **v.** $\begin{pmatrix} 37 & 20 & 99 \\ 47 & 26 & 125 \end{pmatrix}$ **vi. Not Conformable**

AN EASY SHORTCUT

Matrix multiplication is a bizarre and complicated operation that took almost two sessions to completely establish. The idea is to keep practicing. The more you work out, the more you become fluent.

But, if you are having a hard time, there is one method you can follow.

Consider the multiplication,

$$AB = \begin{pmatrix} 1 & 2 \\ 3 & 4 \end{pmatrix} \begin{pmatrix} 5 & 6 \\ 7 & 8 \end{pmatrix}$$

$$2 \times \boxed{2 = 2} \times 2$$

We know that the product has an order 2×2.

The first step is to write the product matrix C in the subscript notation.

$$AB = \begin{pmatrix} 1 & 2 \\ 3 & 4 \end{pmatrix} \begin{pmatrix} 5 & 6 \\ 7 & 8 \end{pmatrix} = \begin{pmatrix} c_{11} & c_{12} \\ c_{21} & c_{22} \end{pmatrix} = C$$

$$2 \times \boxed{2 = 2} \times 2 \qquad\qquad 2 \times 2$$

Now, the subscript of each element in C denotes the row and column of A and B that should be multiplied and added in order to get that element.

For instance, the element c_{11} denotes the first row of A and the first column of B.

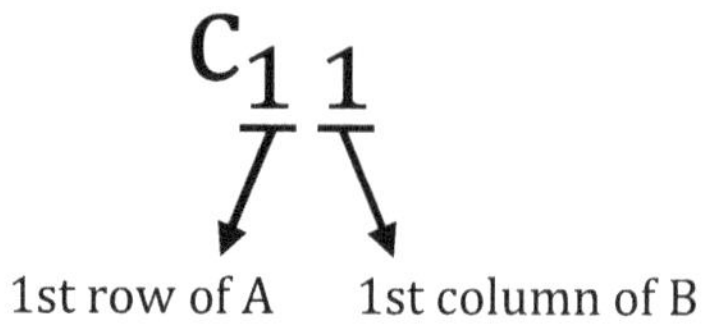

$$c_{1\,1}$$

1st row of A   1st column of B

$$\begin{pmatrix} \boxed{1 \quad 2} \\ 3 \quad 4 \end{pmatrix} \begin{pmatrix} \boxed{\begin{matrix} 5 \\ 7 \end{matrix}} & \begin{matrix} 6 \\ 8 \end{matrix} \end{pmatrix} = \begin{pmatrix} \boxed{c_{11}} & c_{12} \\ c_{21} & c_{22} \end{pmatrix}$$

i.e.,

$$c_{11} = (1)(5) + (2)(7) = 19$$

The element, c_{12} denotes the first row of A and second column of B.

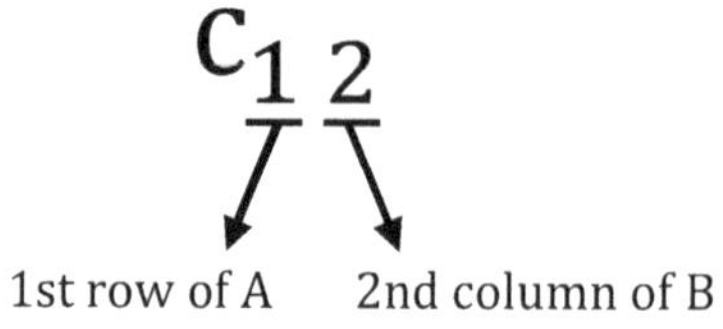

$$c_{1\,2}$$

1st row of A   2nd column of B

$$\begin{pmatrix} \boxed{1 \quad 2} \\ 3 \quad 4 \end{pmatrix} \begin{pmatrix} \begin{matrix} 5 \\ 7 \end{matrix} & \boxed{\begin{matrix} 6 \\ 8 \end{matrix}} \end{pmatrix} = \begin{pmatrix} c_{11} & \boxed{c_{12}} \\ c_{21} & c_{22} \end{pmatrix}$$

i.e.,

$$c_{12} = (1)(6) + (2)(8) = 22$$

Similarly, we get,

$$c_{21} = (3)(5) + (4)(7) = 43$$

and,

$$c_{22} = (3)(6) + (4)(8) = 50$$

Therefore,

$$\begin{pmatrix} 1 & 2 \\ 3 & 4 \end{pmatrix}\begin{pmatrix} 5 & 6 \\ 7 & 8 \end{pmatrix} = \begin{pmatrix} 19 & 22 \\ 43 & 50 \end{pmatrix}$$

This method can used to calculate the product of a column vector and a row vector with the column vector to the left of the row vector. That is if,

$$A = \begin{pmatrix} 1 \\ 2 \\ 3 \end{pmatrix} \text{ and } B = \begin{pmatrix} 4 & 5 & 6 \end{pmatrix}$$

Then,

$$AB = \begin{pmatrix} 1 \\ 2 \\ 3 \end{pmatrix}\begin{pmatrix} 4 & 5 & 6 \end{pmatrix} = \begin{pmatrix} c_{11} & c_{12} & c_{13} \\ c_{21} & c_{22} & c_{23} \\ c_{31} & c_{32} & c_{33} \end{pmatrix} = C$$

$$3 \times \boxed{1 = 1} \times 3 \qquad\qquad 3 \times 3$$

Where,

$$\begin{array}{lll} c_{11} = (1)(4) = 4 & c_{12} = (1)(5) = 5 & c_{13} = (1)(6) = 6 \\ c_{21} = (2)(4) = 8 & c_{22} = (2)(5) = 10 & c_{23} = (2)(6) = 12 \\ c_{31} = (3)(4) = 12 & c_{32} = (3)(5) = 15 & c_{33} = (3)(6) = 18 \end{array}$$

Therefore,

$$\begin{pmatrix} 1 \\ 2 \\ 3 \end{pmatrix}\begin{pmatrix} 4 & 5 & 6 \end{pmatrix} = \begin{pmatrix} 4 & 5 & 6 \\ 8 & 10 & 12 \\ 12 & 15 & 18 \end{pmatrix}$$

Problem 15: If $A = \begin{pmatrix} 1 \\ 0 \\ -1 \end{pmatrix}$; $B = \begin{pmatrix} 1 & 15 & 4 \end{pmatrix}$; $C = \begin{pmatrix} 2 & 3 & 5 \\ -1 & 8 & 9 \end{pmatrix}$ **and** $D = \begin{pmatrix} 1 & 8 \\ -4 & -7 \\ 0 & 6 \end{pmatrix}$, **find:**

i. AB **ii.** BA **iii.** AC **iv.** CA **v.** CB **vi.** CD **vii.** DC

Ans:

i. $\begin{pmatrix} 1 & 15 & 4 \\ 0 & 0 & 0 \\ -1 & -15 & -4 \end{pmatrix}$ **ii.** (-3) **iii.** Not conformable **iv.** $\begin{pmatrix} -3 \\ -10 \end{pmatrix}$

v. Not conformable **vi.** $\begin{pmatrix} -10 & 25 \\ -33 & -10 \end{pmatrix}$ **vii.** $\begin{pmatrix} -6 & 67 & 77 \\ -1 & -68 & -83 \\ -6 & 48 & 54 \end{pmatrix}$

Day Three

Session One

Problem session

Session Two

Row-echelon form

Elementary row operations

Reducing a matrix into an echelon

Session Three

Inverse

Finding inverse using row operations

Meditation is like a gym in which you develop the powerful mental muscle of calm and insight.

- Ajahn Brahm

SESSION ONE

PROBLEM SESSION

Problem solving is a crucial part of mathematical training. The more you work on problems, the more you become an expert. So, let's dedicate this session to solve some problems. But first take a moment to meditate and recollect what we have learned yesterday.

Recall the equality of two matrices, addition of two matrices, scalar multiplication and matrix multiplication.

Example 2: If,

$$A = \begin{pmatrix} 2 & 1 \\ -1 & 3 \end{pmatrix}$$

find A^2.

Solution: The square of a matrix is its product with itself, i.e.

$$A^2 = (A)(A) = \begin{pmatrix} 2 & 1 \\ -1 & 3 \end{pmatrix}\begin{pmatrix} 2 & 1 \\ -1 & 3 \end{pmatrix} = \begin{pmatrix} 3 & 5 \\ -5 & 8 \end{pmatrix}$$

Problem 16: If

$$A = \begin{pmatrix} 1 & 0 & 0 \\ 0 & 0 & 1 \\ 0 & 1 & 0 \end{pmatrix},$$

prove that $A^2 = I$, where I is the 3×3 identity matrix.

$$I = \begin{pmatrix} 1 & 0 & 0 \\ 0 & 1 & 0 \\ 0 & 0 & 1 \end{pmatrix}$$

Problem 17: If $A = \begin{pmatrix} 1 & 1/3 \\ x & y \end{pmatrix}$, find x and y such that $A^2 = 0$.

Where, 0 is the 2×2 null matrix.

$$0 = \begin{pmatrix} 0 & 0 \\ 0 & 0 \end{pmatrix}$$

Ans: $x = -3$; $y = -1$

Note: If A was a number, $A^2 = 0$ clearly means that $A = 0$. But here the elements of A are not zero. This shows that matrices and numbers obey different rules of algebra.

Problem 18: If $A = \begin{pmatrix} 1 & 2 \\ 4 & 3 \end{pmatrix}$, prove that $A^2 - 4A - 5I = 0$.

Where, I is the 2×2 identity matrix and 0 is the 2×2 null matrix.

$$I = \begin{pmatrix} 1 & 0 \\ 0 & 1 \end{pmatrix} \text{ and } 0 = \begin{pmatrix} 0 & 0 \\ 0 & 0 \end{pmatrix}$$

Problem 19: If $A = \begin{pmatrix} 1 & 2 \\ 2 & 3 \end{pmatrix}$ and $A^2 - xA - I = 0$, find x.

Ans: $x = 4$

Problem 20: If $A = \begin{pmatrix} 2 & -2 & -4 \\ -1 & 3 & 4 \\ 1 & -2 & -3 \end{pmatrix}$, prove that $A^2 = A$.

Problem 21: If $A = \begin{pmatrix} 7 & 0 & 3 \\ 4 & -8 & 2 \end{pmatrix}$ and $B = \begin{pmatrix} 4 & -2 \\ 6 & 7 \\ 1 & -1 \end{pmatrix}$, prove that $(AB)^T = B^T A^T$.

The multiplication of a matrix by an identity matrix always results in the matrix itself. That is,

$$\mathbf{IA} = \mathbf{AI} = \mathbf{A}$$

where, '**A**' is any square matrix.

The identity matrix is the matrix analogue of the number '1', since the multiplication of a number by '1' always results in the number itself. i.e.,

$$(1)(a) = (a)(1) = (a)$$

where, 'a' is any number.

If **A** is of the order m × n.

In the product **IA, I** should have an order m × m.

$$\underset{m \times \boxed{m = m} \times n}{(\mathrm{I})\,(\mathrm{A})} = \underset{m \times n}{(\mathrm{A})}$$

And in the product **AI, I** should have an order n × n.

$$\underset{m \times \boxed{n = n} \times n}{(\mathrm{A})\,(\mathrm{I})} = \underset{m \times n}{(\mathrm{A})}$$

<u>Example 3</u>: If A = $\begin{pmatrix} 2 & 1 & 5 \\ 7 & 0 & 8 \end{pmatrix}$, find IA and AI.

<u>Solution</u>: Since A has an order 2 × 3, in the product IA, I should have an order 2 × 2.

Therefore,

$$\mathbf{IA} = \begin{pmatrix} 1 & 0 \\ 0 & 1 \end{pmatrix}\begin{pmatrix} 2 & 1 & 5 \\ 7 & 0 & 8 \end{pmatrix} = \begin{pmatrix} 2 & 1 & 5 \\ 7 & 0 & 8 \end{pmatrix}$$

And in the product AI, I should have an order 3 × 3.

Therefore,

$$\mathbf{AI} = \begin{pmatrix} 2 & 1 & 5 \\ 7 & 0 & 8 \end{pmatrix}\begin{pmatrix} 1 & 0 & 0 \\ 0 & 1 & 0 \\ 0 & 0 & 1 \end{pmatrix} = \begin{pmatrix} 2 & 1 & 5 \\ 7 & 0 & 8 \end{pmatrix}$$

Problem 22: If I is the identity matrix, show that

$$\mathbf{IA = AI = A}$$

for,

i. $A = \begin{pmatrix} 2 & -2 & -4 \\ -1 & 3 & 4 \\ 1 & -2 & -3 \end{pmatrix}$ **ii.** $A = \begin{pmatrix} 0 & 1 \\ 1 & 0 \end{pmatrix}$ **iii.** $A = \begin{pmatrix} 1 & -1 & 8 \end{pmatrix}$

iv. $A = \begin{pmatrix} 7 & 8 \\ 14 & -1 \\ -2 & 6 \end{pmatrix}$ **v.** $A = \begin{pmatrix} 2 \\ 7 \\ 3 \end{pmatrix}$

SESSION TWO

ROW-ECHELON FORM

A matrix is said to be in **row-echelon** form if it satisfies the following conditions.

i. If the matrix contains any zero rows (rows with only zeroes), they should be at the bottom. For example,

$$\begin{pmatrix} 1 & 2 & -8 \\ 0 & 0 & 0 \end{pmatrix};\ \begin{pmatrix} 1 & 0 \\ 0 & 0 \end{pmatrix};\ \begin{pmatrix} 1 & 2 & 3 \\ 0 & 0 & 0 \\ 0 & 0 & 0 \end{pmatrix} \text{ etc.}$$

The matrix,

$$\begin{pmatrix} 0 & 0 \\ 2 & 1 \end{pmatrix}$$

is not in row-echelon form, since the row containing only zeroes is not at the bottom.

ii. The left most element in each non-zero row should be equal to '1' (called the leading one). For example,

$$\begin{pmatrix} \boxed{1} & 2 & -8 \\ 0 & \boxed{1} & 5 \end{pmatrix};\ \begin{pmatrix} \boxed{1} & 8 \\ 0 & \boxed{1} \end{pmatrix};\ \begin{pmatrix} \boxed{1} & 2 & 3 \\ 0 & \boxed{1} & 2 \\ 0 & 0 & 0 \end{pmatrix} \text{ etc}$$

The matrix,

$$\begin{pmatrix} \boxed{2} & 3 \\ 0 & 0 \end{pmatrix}$$

is not in row-echelon form, since the left most element in the non-zero row is not a leading one.

iii. All the elements below the leading one should be zero.

For example,

$$\begin{pmatrix} 1 & 2 & 5 \\ \boxed{0} & 1 & 8 \\ \boxed{0} & \boxed{0} & 0 \end{pmatrix} ; \begin{pmatrix} 1 & 8 \\ \boxed{0} & 1 \end{pmatrix} ; \begin{pmatrix} 1 & -5 & 5 \\ \boxed{0} & 1 & 3 \\ \boxed{0} & \boxed{0} & 1 \end{pmatrix} \text{ etc}$$

The matrix,

$$\begin{pmatrix} 1 & 2 & 3 \\ \boxed{1} & 8 & 0 \\ \boxed{0} & 0 & 1 \end{pmatrix}$$

is not row-echelon form, since all the elements below the leading one of the first row are not zero.

Examples of matrices in row echelon form are,

$$\begin{pmatrix} 1 & 0 & 8 & 4 \\ 0 & 1 & 2 & 5 \\ 0 & 0 & 1 & 9 \end{pmatrix} ; \begin{pmatrix} 1 & 8 & 3 \\ 0 & 1 & 4 \\ 0 & 0 & 1 \end{pmatrix} ; \begin{pmatrix} 1 & -4 & 15 \\ 0 & 1 & 2 \\ 0 & 0 & 0 \end{pmatrix}$$

<u>Problem 23</u>: Which of the following matrices are not in row-echelon form.

$$\mathbf{A} = \begin{pmatrix} 1 & 7 \\ 0 & 1 \\ 0 & 0 \end{pmatrix} \qquad \mathbf{B} = \begin{pmatrix} 1 & -5 \\ 0 & 0 \end{pmatrix} \qquad \mathbf{C} = \begin{pmatrix} 1 & 4 \\ 0 & -1 \end{pmatrix}$$

$$\mathbf{D} = \begin{pmatrix} 1 & -9 & -10 \\ 0 & 0 & 0 \\ 0 & 0 & 1 \end{pmatrix} \qquad \mathbf{E} = \begin{pmatrix} 1 & 4 \\ 0 & 1 \end{pmatrix} \qquad \mathbf{F} = \begin{pmatrix} 1 & -7 & -7 \\ 0 & 0 & 1 \\ 0 & -1 & 0 \end{pmatrix}$$

<u>Ans</u>: C, D and F

ELEMENTARY ROW OPERATIONS

Using a set of three operations, any matrix can be converted into a row-echelon form. They are known as the elementary row operations. They are:

i. Interchanging the positions of any two rows.

ii. Multiplying any row by a non-zero scalar.

iii. Add to one row, a scalar times another row.

For example,

In the matrix,

$$\begin{pmatrix} 7 & 1 & 2 \\ 5 & 10 & 4 \\ 7 & 8 & 3 \end{pmatrix}$$

The three rows can be named as R_1, R_2 and R_3 respectively.

$$\begin{pmatrix} \boxed{7 \quad 1 \quad 2} \\ \boxed{5 \quad 10 \quad 4} \\ \boxed{7 \quad 8 \quad 3} \end{pmatrix} \begin{matrix} \rightarrow R_1 \\ \rightarrow R_2 \\ \rightarrow R_3 \end{matrix}$$

Interchanging the first and the second row ($R_1 \leftrightarrow R_2$) gives,

$R_1 \leftrightarrow R_2$

$$\begin{pmatrix} 5 & 10 & 4 \\ 7 & 1 & 2 \\ 7 & 8 & 3 \end{pmatrix}$$

Multiplying the second row by the number '2' ($R_2 \longrightarrow 2R_2$) gives,

$R_2 \longrightarrow 2R_2$

$$\begin{pmatrix} 7 & 1 & 2 \\ 10 & 20 & 8 \\ 7 & 8 & 3 \end{pmatrix}$$

And, multiplying the third row by 3 and adding it to the first row ($R_1 \longrightarrow R_1 + 3R_3$) gives,

$R_1 \longrightarrow R_1 + 3R_3$

$$\begin{pmatrix} 28 & 25 & 11 \\ 5 & 10 & 4 \\ 7 & 8 & 3 \end{pmatrix}$$

To reduce errors while doing the row operations, first write down the rows that do not have any change, and then write the row that has been changed by the operation.

Problem 24: Execute the following row operations on the matrix $\begin{pmatrix} 1 & 2 & 3 \\ 4 & 5 & 6 \\ 7 & 8 & 9 \end{pmatrix}$ distinctively.

i. **Multiply the first row by $\frac{1}{7}$.**

   $(R_1 \to \frac{1}{7} R_1)$

ii. **Interchange second and third rows.**

   $(R_2 \leftrightarrow R_3)$

iii. **Multiply the second row with -1 and add it to the third.**

   $(R_3 \to R_3 + (-1)R_2)$

Ans:

i. $\begin{pmatrix} 1/7 & 2/7 & 3/7 \\ 4 & 5 & 6 \\ 7 & 8 & 9 \end{pmatrix}$ ii. $\begin{pmatrix} 1 & 2 & 3 \\ 7 & 8 & 9 \\ 4 & 5 & 6 \end{pmatrix}$ iii. $\begin{pmatrix} 1 & 2 & 3 \\ 4 & 5 & 6 \\ 3 & 3 & 3 \end{pmatrix}$

REDUCING A MATRIX INTO AN ECHELON FORM

Using a sequential employment of the three elementary row operations, any matrix can be converted into a row-echelon form. See the example below.

Example 4: Convert the following matrices into row-echelon forms using elementary row operations.

i. $\begin{pmatrix} 9 & 12 \\ 8 & 1 \end{pmatrix}$ ii. $\begin{pmatrix} 2 & -1 & 8 \\ 0 & 0 & 0 \\ -7 & 15 & 4 \end{pmatrix}$

Solution:

i. Consider the matrix

$$\begin{pmatrix} 9 & 12 \\ 8 & 1 \end{pmatrix}$$

Here, the left most element of the first row is not a leading one. Therefore, we employ the operation, $R_1 \rightarrow \frac{1}{9} R_1$,which results in,

$$\begin{pmatrix} 1 & 4/3 \\ 8 & 1 \end{pmatrix}$$

Now, the next operation must be done on the new matrix, not on the original one.

In the new matrix, the element below the leading one is not zero. Therefore, we employ the operation $R_2 \rightarrow R_2 + (-8)R_1$ on the new matrix, resulting in,

$$\begin{pmatrix} 1 & 4/3 \\ 0 & -29/3 \end{pmatrix}$$

To make the non-zero element of the second row a leading one, we employ the operation $R_2 \rightarrow -\frac{3}{29} R_2$,which gives the row-echelon form:

$$\begin{pmatrix} 1 & 4/3 \\ 0 & 1 \end{pmatrix}$$

To summarize,

$$\begin{pmatrix} 9 & 12 \\ 8 & 1 \end{pmatrix} \xrightarrow{R_1 \rightarrow \frac{1}{9} R_1} \begin{pmatrix} 1 & 4/3 \\ 8 & 1 \end{pmatrix} \xrightarrow{R_2 \rightarrow R_2 + (-8)R_1} \begin{pmatrix} 1 & 4/3 \\ 0 & -29/3 \end{pmatrix} \xrightarrow{R_2 \rightarrow -\frac{3}{29} R_2} \begin{pmatrix} 1 & 4/3 \\ 0 & 1 \end{pmatrix}$$

ii. Consider the matrix

$$\begin{pmatrix} 2 & -1 & 8 \\ 0 & 0 & 0 \\ -7 & 15 & 4 \end{pmatrix}$$

Employing the row operations consecutively,

$R_2 \leftrightarrow R_3$

$$\begin{pmatrix} 2 & -1 & 8 \\ -7 & 15 & 4 \\ 0 & 0 & 0 \end{pmatrix}$$

$R_1 \longrightarrow \frac{1}{2} R_1$

$$\begin{pmatrix} 1 & -1/2 & 4 \\ -7 & 15 & 4 \\ 0 & 0 & 0 \end{pmatrix}$$

$R_2 \longrightarrow R_2 + 7R_1$

$$\begin{pmatrix} 1 & -1/2 & 4 \\ 0 & 23/2 & 32 \\ 0 & 0 & 0 \end{pmatrix}$$

$R_2 \longrightarrow -\frac{2}{37} R_2$

$$\begin{pmatrix} 1 & -1/2 & 4 \\ 0 & 1 & 64/23 \\ 0 & 0 & 0 \end{pmatrix}$$

This is the row-echelon form

Problem 25: Reduce the following matrices into row-echelon form:

i. $\begin{pmatrix} 1 & 2 & 3 \\ 4 & 5 & 6 \\ 7 & 8 & 9 \end{pmatrix}$ ii. $\begin{pmatrix} 9 & 2 & 6 \\ 8 & -4 & 5 \\ 7 & 5 & 7 \end{pmatrix}$ iii. $\begin{pmatrix} 7 & 2 & 11 \\ 6 & 3 & -3 \\ 5 & 0 & 0 \end{pmatrix}$

Ans:

i. $\begin{pmatrix} 1 & 2 & 3 \\ 0 & 1 & 2 \\ 0 & 0 & 0 \end{pmatrix}$ ii. $\begin{pmatrix} 1 & 2/9 & 2/3 \\ 0 & 1 & 3/52 \\ 0 & 0 & 1 \end{pmatrix}$ iii. $\begin{pmatrix} 1 & 2/7 & 11/7 \\ 0 & 1 & -87/9 \\ 0 & 0 & 1 \end{pmatrix}$

Reducing a matrix into its row-echelon form is a crucial part of solving linear equations, which we will learn in the upcoming chapters. But for now, let's take a break.

In the next session, we shall use the row operations to find the inverse of a matrix.

SESSION THREE

INVERSE

In session one, we have seen that the product of a matrix with an identity matrix always yields itself. That is,

$$\mathbf{IA} = \mathbf{AI} = \mathbf{A}$$

where, **A** is any matrix.

This is similar to the multiplication of a number by the number '1'. For instance,

$$(1)(2) = (2)(1) = 2$$

or,

$$(1)(3) = (3)(1) = 3$$

Hence, the number '1' is said to be the multiplicative identity for numbers.

Now, a number is said to be the **inverse** of another number **if their product always yields the identity**. For instance, the number $\frac{1}{2}$ is said to be the inverse of the number 2, since their product always yields the identity. That is,

$$(2)\left(\frac{1}{2}\right) = \left(\frac{1}{2}\right)(2) = 1$$

or,

The number $\frac{1}{3}$ is said to be the inverse of the number 3, since their product always yields the identity. That is,

$$(3)\left(\frac{1}{3}\right) = \left(\frac{1}{3}\right)(3) = 1$$

Similarly, A matrix **A** is said to have an **inverse** $\mathbf{A}^{-1}$("A-inverse"), if their product always yields the identity matrix. That is,

$$\mathbf{AA}^{-1} = \mathbf{A}^{-1}\mathbf{A} = \mathbf{I}$$

The commutativity of **A** and $\mathbf{A}^{-1}$ shows that they both are square matrices of the same order.

$$\underset{n\times \boxed{n\ =\ n}\times n}{(A)\,(A^{-1})} = \underset{n\times \boxed{n\ =\ n}\times n}{(A^{-1})\,(A)} = \underset{n\times n}{(I)}$$

Example 5: Show that the matrix $\begin{pmatrix}4 & 3\\3 & 2\end{pmatrix}$ is the inverse of $\begin{pmatrix}-2 & 3\\3 & -4\end{pmatrix}$.

Solution: Let $A = \begin{pmatrix}4 & 3\\3 & 2\end{pmatrix}$ and $B = \begin{pmatrix}-2 & 3\\3 & -4\end{pmatrix}$.

Thus,

$$AB = \begin{pmatrix}4 & 3\\3 & 2\end{pmatrix}\begin{pmatrix}-2 & 3\\3 & -4\end{pmatrix} = \begin{pmatrix}1 & 0\\0 & 1\end{pmatrix} = I$$

and,

$$BA = \begin{pmatrix}-2 & 3\\3 & -4\end{pmatrix}\begin{pmatrix}4 & 3\\3 & 2\end{pmatrix} = \begin{pmatrix}1 & 0\\0 & 1\end{pmatrix} = I$$

i.e.,

$$AB = BA = I$$

On comparison with the equation

$$AA^{-1} = A^{-1}A = I$$

we find,

$$B = A^{-1}$$

Problem 26: Show that the matrix $\begin{pmatrix}-1/2 & -5/4 & 3/4\\1/2 & 3/4 & -1/4\\-1/2 & 1/4 & 1/4\end{pmatrix}$ is the inverse of $\begin{pmatrix}1 & 2 & -1\\0 & 1 & 1\\2 & 3 & 1\end{pmatrix}$.

FINDING INVERSE USING ROW OPERATIONS

There are many ways to find the inverse of a square matrix. An efficient way is by employing the elementary row operations. The method is similar to the row-echelon reduction.

Let **A** be a square matrix whose inverse is to be found. The first step is to form the block matrix.

$$\mathbf{(A\,|\,I)}$$

Where, the matrix is joined with an identity matrix of the same order separated by a vertical line.

For example, if

$$\mathbf{A} = \begin{pmatrix} 4 & 3 \\ 3 & 2 \end{pmatrix}$$

Then,

$$(\mathbf{A} \mid \mathbf{I}) = \left(\begin{array}{cc|cc} 4 & 3 & 1 & 0 \\ 3 & 2 & 0 & 1 \end{array}\right)$$

This new matrix is called a **block matrix**, since it is **partitioned into two blocks**, **A** and **I**.

Now, perform the elementary row operations on the entire block matrix to transform **A** into an identity matrix. That is,

$$\left(\begin{array}{cc|cc} 4 & 3 & 1 & 0 \\ 3 & 2 & 0 & 1 \end{array}\right)$$

$$R_1 \longrightarrow \frac{1}{4} R_1$$

$$\left(\begin{array}{cc|cc} 1 & 3/4 & 1/4 & 0 \\ 3 & 2 & 0 & 1 \end{array}\right)$$

$$R_2 \longrightarrow R_2 - 3R_1$$

$$\left(\begin{array}{cc|cc} 1 & 3/4 & 1/4 & 0 \\ 0 & -1/4 & -3/4 & 1 \end{array}\right)$$

$$R_2 \longrightarrow -4R_2$$

$$\left(\begin{array}{cc|cc} 1 & 3/4 & 1/4 & 0 \\ 0 & 1 & 3 & -4 \end{array}\right)$$

$$R_1 \longrightarrow R_1 - \frac{3}{4} R_2$$

$$\left(\begin{array}{cc|cc} 1 & 0 & -2 & 3 \\ 0 & 1 & 3 & -4 \end{array}\right)$$

When **A** is transformed into the identity matrix, the matrix on the right block will be the inverse of **A**. That is,

$$\mathbf{A}^{-1} = \begin{pmatrix} -2 & 3 \\ 3 & -4 \end{pmatrix}$$

This can be verified by the equation $\mathbf{AA}^{-1} = \mathbf{A}^{-1}\mathbf{A} = \mathbf{I}$ (See example:5).

To summarize,

$$(\mathbf{A} \mid \mathbf{I}) \xrightarrow{\text{row opetations}} (\mathbf{I} \mid \mathbf{A}^{-1})$$

See the following example,

Example 6: Find the inverse of the matrix $A = \begin{pmatrix} 1 & 2 & -1 \\ 0 & 1 & 1 \\ 2 & 3 & 1 \end{pmatrix}$ using elementary row operations.

Solution: Write the block matrix,

$$(A \mid I) = \left(\begin{array}{ccc|ccc} 1 & 2 & -1 & 1 & 0 & 0 \\ 0 & 1 & 1 & 0 & 1 & 0 \\ 2 & 3 & 1 & 0 & 0 & 1 \end{array}\right)$$

$R_3 \longrightarrow R_3 - 2R_1$

$$\left(\begin{array}{ccc|ccc} 1 & 2 & -1 & 1 & 0 & 0 \\ 0 & 1 & 1 & 0 & 1 & 0 \\ 0 & -1 & 3 & -2 & 0 & 1 \end{array}\right)$$

$R_1 \longrightarrow R_1 + 2R_3$

$$\left(\begin{array}{ccc|ccc} 1 & 0 & 5 & -3 & 0 & 2 \\ 0 & 1 & 1 & 0 & 1 & 0 \\ 0 & -1 & 3 & -2 & 0 & 1 \end{array}\right)$$

$R_3 \longrightarrow R_3 + R_2$

$$\left(\begin{array}{ccc|ccc} 1 & 0 & 5 & -3 & 0 & 2 \\ 0 & 1 & 1 & 0 & 1 & 0 \\ 0 & 0 & 4 & -2 & 1 & 1 \end{array}\right)$$

$R_3 \longrightarrow \frac{1}{4}R_3$

$$\left(\begin{array}{ccc|ccc} 1 & 0 & 5 & -3 & 0 & 2 \\ 0 & 1 & 1 & 0 & 1 & 0 \\ 0 & 0 & 1 & -1/2 & 1/4 & 1/4 \end{array}\right)$$

$R_2 \longrightarrow R_2 - R_3$

$$\left(\begin{array}{ccc|ccc} 1 & 0 & 5 & -3 & 0 & 2 \\ 0 & 1 & 0 & 1/2 & 3/4 & -1/4 \\ 0 & 0 & 1 & -1/2 & 1/4 & 1/4 \end{array}\right)$$

$R_1 \longrightarrow R_1 - 5R_3$

$$\left(\begin{array}{ccc|ccc} 1 & 0 & 0 & -1/2 & -5/4 & 3/4 \\ 0 & 1 & 0 & 1/2 & 3/4 & -1/4 \\ 0 & 0 & 1 & -1/2 & 1/4 & 1/4 \end{array}\right) = (I \mid A^{-1})$$

Therefore,

$$A^{-1} = \begin{pmatrix} -1/2 & -5/4 & 3/4 \\ 1/2 & 3/4 & -1/4 \\ -1/2 & 1/4 & 1/4 \end{pmatrix}$$

This can be verified by the equation $AA^{-1} = A^{-1}A = I$ (see problem:26).

A matrix is said to be **singular** if it **does not have an inverse**. A singular matrix is identified if we encounter a zero row during the row transformations. See the example,

Example 7: Show that the matrix $\begin{pmatrix} 2 & 4 \\ 3 & 6 \end{pmatrix}$ does not have an inverse.

Solution: Let,

$$A = \begin{pmatrix} 2 & 4 \\ 3 & 6 \end{pmatrix}$$

Then,

$$(A \mid I) = \left(\begin{array}{cc|cc} 2 & 4 & 1 & 0 \\ 3 & 6 & 0 & 1 \end{array}\right)$$

$R_1 \longrightarrow \frac{1}{2} R_1$

$$\left(\begin{array}{cc|cc} 1 & 2 & 1/2 & 0 \\ 3 & 6 & 0 & 1 \end{array}\right)$$

$R_2 \longrightarrow R_2 - 3R_1$

$$\left(\begin{array}{cc|cc} 1 & 2 & 1/2 & 0 \\ 0 & 0 & -3/2 & 1 \end{array}\right)$$

The appearance of a zero row during the row transformation of A shows that A is singular.

Problem 27: Find the inverses of the following matrices.

i. $\begin{pmatrix} 2 & 0 \\ 1 & 1 \end{pmatrix}$ ii. $\begin{pmatrix} 1 & 0 & 2 \\ 2 & -1 & 3 \\ 4 & 1 & 8 \end{pmatrix}$ iii. $\begin{pmatrix} 2 & 3 \\ 2 & 1 \end{pmatrix}$

iv. $\begin{pmatrix} 3 & 0 & 0 \\ -1 & 2 & 0 \\ 1 & 1 & 0 \end{pmatrix}$ v. $\begin{pmatrix} 9 & 6 \\ 3 & 2 \end{pmatrix}$

Ans:

i. $\begin{pmatrix} 1/2 & 0 \\ -1/2 & 1 \end{pmatrix}$ ii. $\begin{pmatrix} -11 & 2 & 2 \\ -4 & 0 & 1 \\ 6 & -1 & -1 \end{pmatrix}$ iii. $\begin{pmatrix} -1/4 & 3/4 \\ 1/2 & -1/2 \end{pmatrix}$

iv. No Inverse v. No Inverse

That's it, we are done with the basics. In the upcoming chapters, we will reinvent the concept of matrices in the light of linear equations. But for now, let's wrap it up.

Before ending the session, remember to meditate and recall everything. If you could not complete the sessions in one day, don't worry, it's ok. Tomorrow, return to the same session. Try to create a good mental picture and make flash cards for future revisions.

Day Four

Session One

Linear equations

Session Two

Existence of exact solution

Linear systems

Special cases of linear systems

Session Three

Connecting matrices and linear equations

A man who dares to waste one hour of time has not discovered the value of life.

- Charles Darwin

SESSION ONE

LINEAR EQUATIONS

An equation is a statement that asserts an equality between two mathematical expressions. For instance, the equation

$$\mathbf{2x + 3 = 1}$$

equates the expressions $2x + 3$ and 1, where 1, 2 and 3 are numbers,

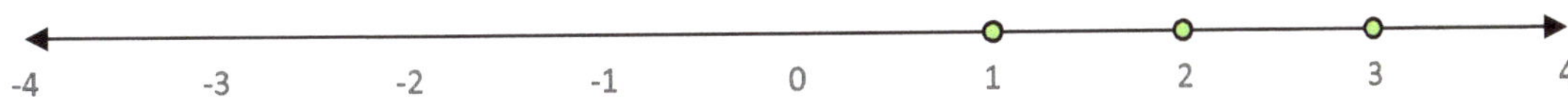

and 'x' is a variable whose value is unknown.

To solve an equation is to find the value of its variable. It is done using the basic rules of algebra. If you add a term to the Left Hand Side (LHS), you do the same to the Right Hand Side (RHS). That is, if,

$$a = b$$

then,

$$a + c = b + c$$

Same goes with subtraction,

$$a - c = b - c$$

Multiplication,

$$a(c) = b(c)$$

and division,

$$\frac{a}{c} = \frac{b}{c}$$

Using these rules, the equation $2x + 3 = 1$ can be solved as,

$$2x + 3 - 3 = 1 - 3$$

or

$$2x = -2$$

$$\frac{2x}{2} = -\frac{2}{2}$$

or

$$x = -1$$

Thus the equation is solved with the solution,

$$x = -1$$

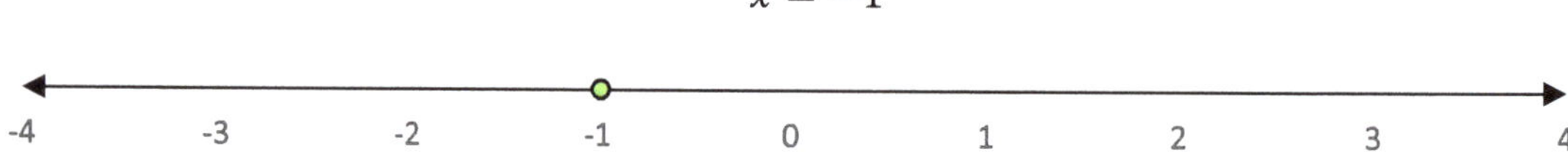

The solution of an equation always respects its equality. This is verified by substituting the solution back into the equation. That is,

$$2(-1) + 3 = 1$$

or

$$-2 + 3 = 1$$

or

$$1 = 1$$

Since, $1 = 1$ is a true statement of equality, we conclude that $x = -1$ is a solution.

Also, it is easy to find that no other value of 'x' can result in an equality. For instance, if $x = 1$, we find,

$$2(1) + 3 = 5 \neq 1$$

That is $x = 1$ violates the equality, and therefore is not a solution.

Problem 28: Solve the following linear equations and verify the solutions. Also plot the solutions on a number line.

i. $\mathbf{5x + 3 = 8}$ **ii.** $\mathbf{2 = 3y}$

iii. $\mathbf{x + 2x = 6}$ **iv.** $\mathbf{\frac{1}{x} = 2}$

Ans:

i. $\mathbf{x = 1}$ **ii.** $\mathbf{y = \frac{2}{3}}$ **iii.** $\mathbf{x = 2}$ **iv.** $\mathbf{x = \frac{1}{2}}$

Now consider an equation having two variables 'x' and 'y'. For instance,

$$\mathbf{2x + 3y = 1}$$

Upon solving for 'x',

$$2x + 3y - 3y = 1 - 3y$$

or

$$2x = 1 - 3y$$

$$\frac{2x}{2} = \frac{1 - 3y}{2}$$

or

$$x = \frac{1}{2} - \frac{3}{2}y$$

Here, we find that 'x' is a function of 'y'. Meaning, for each value of y, there exists an 'x', defined by the equation.

$$x = \frac{1}{2} - \frac{3}{2}y$$

That is, if

$$y = 0$$

we get,

$$x = \frac{1}{2} - \frac{3}{2}(0) = \frac{1}{2}$$

Thus, if we write 'x' and 'y' as an ordered pair (x, y). We get,

$$(x, y) = \left(\frac{1}{2}, 0\right)$$

Similarly, if

$$y = 1$$

we get,

$$x = \frac{1}{2} - \frac{3}{2}(1) = -1$$

Therefore,

$$(x, y) = (-1,1)$$

Since 'y' can have any real number as its value, the number of ordered pairs (x, y) is actually infinite.

A few of them are listed in the table below.

y	$x = \frac{1}{2} - \frac{3}{2}y$	(x, y)
$\vdots$	$\vdots$	$\vdots$
-3	$x = \frac{1}{2} - \frac{3}{2}(-3) = 5$	$(5, -3)$
-2	$x = \frac{1}{2} - \frac{3}{2}(-2) = 3.5$	$(3.5, -2)$
-1	$x = \frac{1}{2} - \frac{3}{2}(-1) = 2$	$(2, -1)$
0	$x = \frac{1}{2} - \frac{3}{2}(0) = .5$	$(.5,0)$
1	$x = \frac{1}{2} - \frac{3}{2}(1) = -1$	$(-1,1)$
2	$x = \frac{1}{2} - \frac{3}{2}(2) = -2.5$	$(-2.5,2)$
3	$x = \frac{1}{2} - \frac{3}{2}(3) = -4$	$(-4,3)$
$\vdots$	$\vdots$	$\vdots$

Here, each ordered pair (x, y) respects the equality of the equation $2x + 3y = 1$. For instance, if

$$(x, y) = (5, -3)$$

The equation becomes.

$$2(5) + 3(-3) = 1$$

or

$$1 = 1$$

or if

$$(x, y) = (3.5, -2)$$

The equation becomes,

$$2(3.5) + 3(-2) = 1$$

or

$$1 = 1$$

This shows that each ordered pair (x, y) is a solution of the equation. And, since an infinite number of ordered pairs are possible, the equation $2x + 3y = 1$ has **infinite** number of **solutions.**

The solutions can be represented as the ordered pair,

$$\left(\frac{1}{2} - \frac{3}{2}y, y\right)$$

or as the set of equations,

$$\left.\begin{array}{c} x = \dfrac{1}{2} - \dfrac{3}{2}y \\ y = y \end{array}\right\}$$

The presence of two variables in the equation demands two number lines for plotting the solutions. So we use an X-Y cordinate system. If you are not familiar with a coordinate system, see the box below, if you are, then skip it.

HOW TO WORK WITH CORDINATE SYSTEMS

A coordinate system is a device used to locate and name points in a plane. It consists of two infinitely long perpendicular lines called the 'X' and 'Y' axes, meeting at a single point called the 'origin'. Every point on the axes corresponds to a real number. The origin corresponds to the number zero.

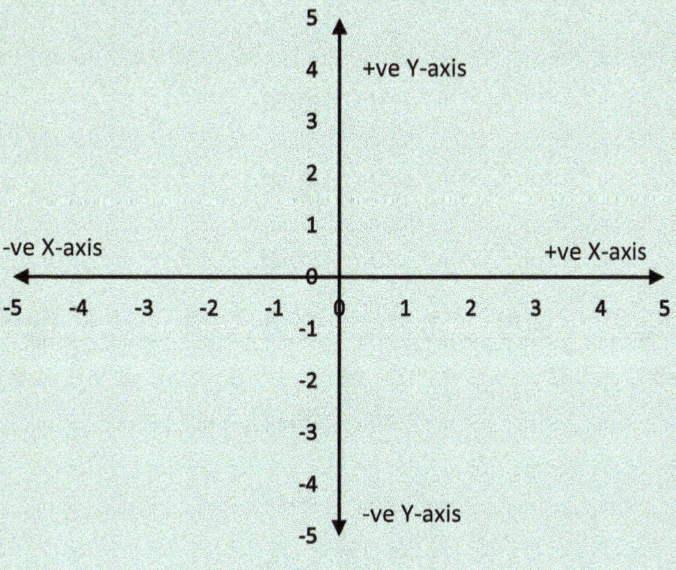

A point is named as a pair of numbers (x, y) called coordinates. The name (x, y) suggests that to reach the point, one has to move 'x' units along the direction of the X-axis, and then 'y' units along the direction of the Y-axis.

For instance, in the figure, the point P is named (3,2), since to reach P, we have to move 3 units along the direction of the positive X-axis and then 2 units along the direction of the positive Y-axis. The number 3 is called the 'x-coordinate' and 2 is called the 'y-coordinate'.

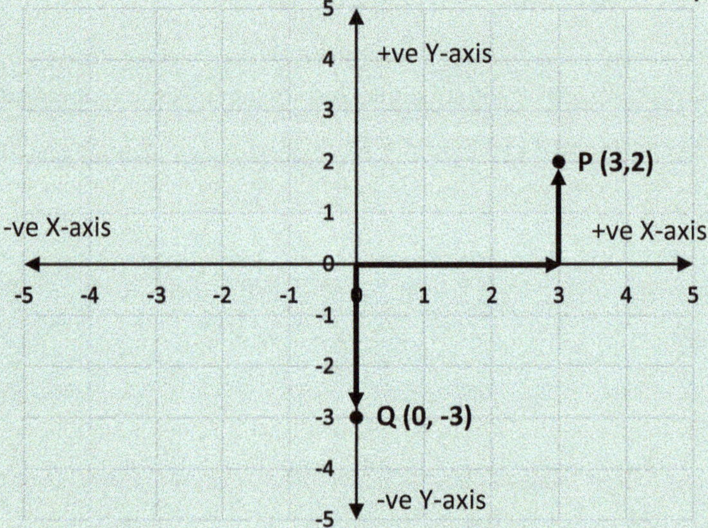

The point Q is named $(0, -3)$, since to reach Q we have to move 0 units along the X-axis and then 3 units along the negative Y-axis. Here 0 is the x-coordinate and -3 is the y-coordinate.

For all points along the X-axis, the y-coordinate is zero, and for all points along the Y-axis, the x-coordinate is zero.

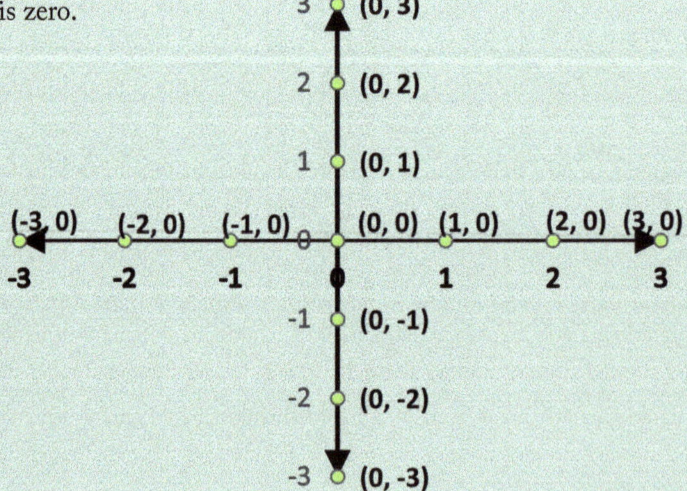

Problem: Draw a coordinate system and locate the following points:

i.	$(0,-1.5)$	**ii.**	$(-4,-3)$	**iii.**	$(1,-2)$	**iv.**	$(6,4)$
v.	$(-4,3)$	**vi.**	$(1,1.5)$				

Upon plotting the solutions (x, y)

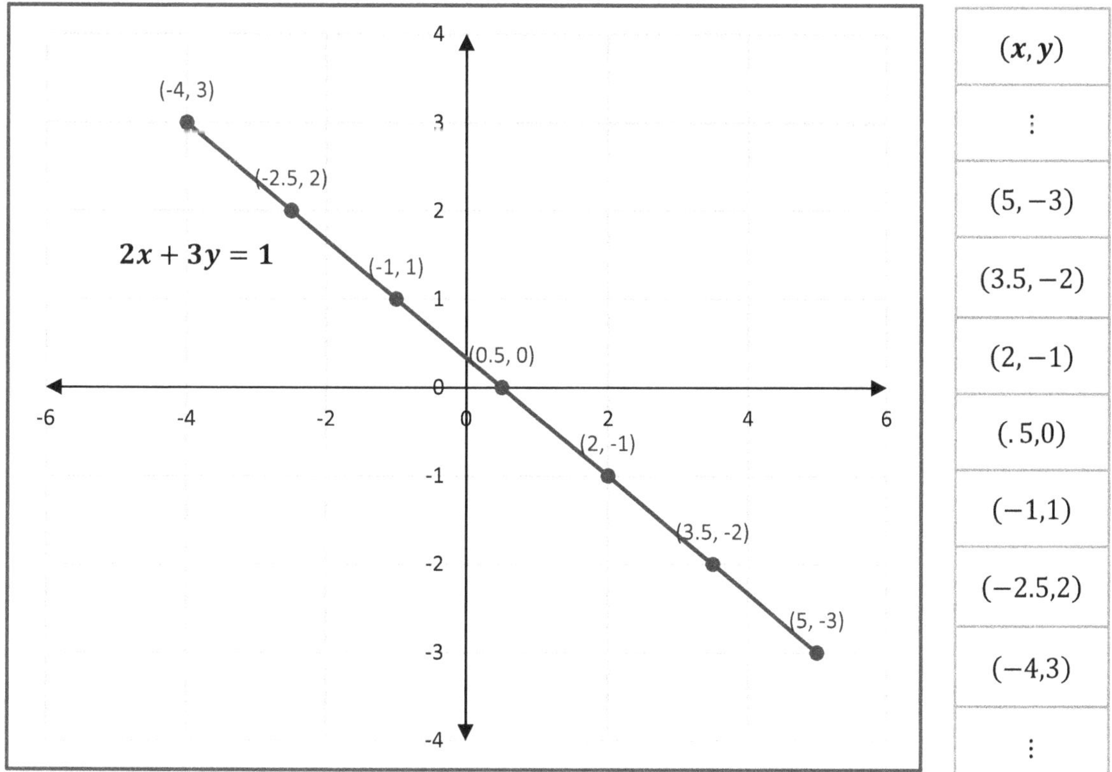

(x, y)
⋮
$(5,-3)$
$(3.5,-2)$
$(2,-1)$
$(.5,0)$
$(-1,1)$
$(-2.5,2)$
$(-4,3)$
⋮

We obtain a **straight line.**

Hence the name **'Linear Equation'**.

Note that the infinite solutions of the linear equation correspond to the infinite points along the straight line written as the ordered pairs (x, y).

Some examples of linear equations having two variables are,

$2x + 5y = 7$
$3x_1 + 7x_2 = 8$
$y - 7x = 10$
$2x = -13$ (Here, the coefficient of 'y' is zero therefore it is equivalent to $2x + 0y = -13$)

$5y = 1$ (This equation is equivalent to $0x + 5y = 1$) etc.

In general, a linear equation of two variables is of the form

$$ax + by = c$$

where 'a', 'b' and 'c' are constants.

A linear equation of three variables is of the form

$$ax + by + cz = d$$

where 'a', 'b', 'c' and 'd' are constants.

And a general linear equation of 'n' variables is of the form

$$a_1x_1 + a_2x_2 + \cdots + a_nx_n = b$$

where 'a_1', 'a_2'... 'a_n' and 'b' are constants.

If $b = 0$, the equation is said to be homogeneous. That is

$$a_1x_1 + a_2x_2 + \cdots + a_nx_n = 0$$

Examples of homogeneous equations are,

$2x + 3y = 0$
$5x + 7y - z = 0$
$2x = 0$
$y = 0$
$15y - \frac{1}{2}x = 0$ etc.

HOW TO PLOT A LINEAR EQUATION

Consider the equation,

$$2x + 3y = 6$$

To plot the equation, find the points where it meets the coordinate axes ("intercepts").

To find the Y- intercept, put $x = 0$ into the equation. That is,

$$2(0) + 3(y) = 6$$

or

$$y = 2$$

This corresponds to the point (0,2).

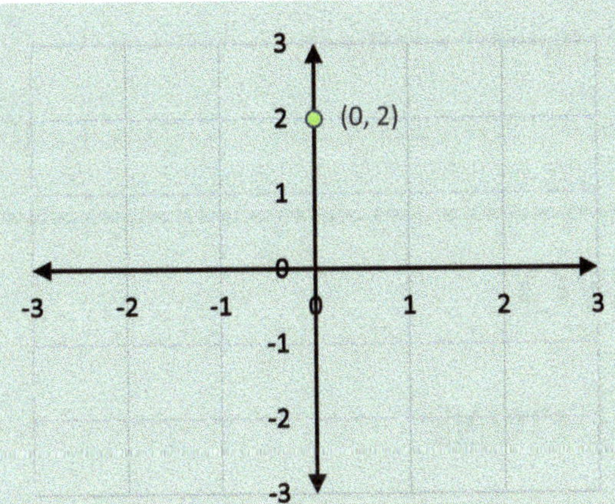

To find the X- intercept, put $y = 0$ into the equation. That is,

$$2(x) + 3(0) = 6$$

or

$$x = 3$$

This corresponds to the point (3,0).

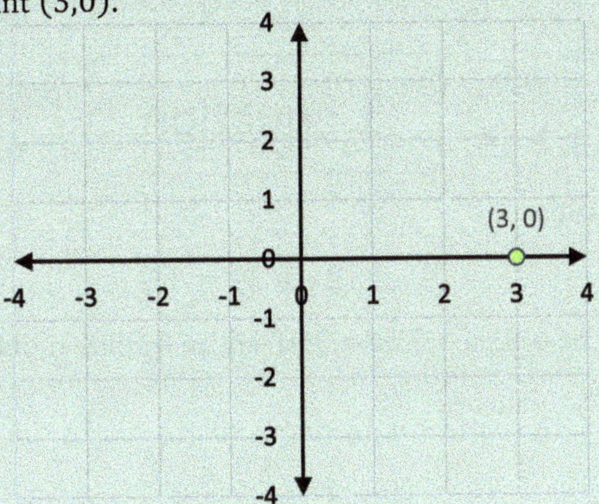

Now, join the two points to obtain the straight line,

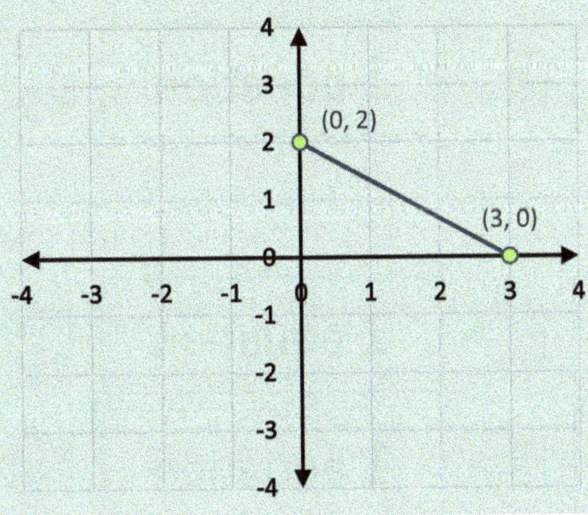

Problem: Plot the following linear equations:

i. $x + y = 1$ **ii.** $x - y = 1$ **iii.** $2x + y = 5$ **iv.** $2x + 4y = 8$

SESSION TWO

NATURE OF SOLUTIONS

Okay, let's recap the previous session. First, we studied the equation

$$2x + 3 = 1$$

having one variable and a unique solution

$$x = -1$$

That is,

1 variable  1 equation  Unique solution

Next, we studied the equation

$$2x + 3y = 1$$

with two variables having no unique solution, instead, an infinite number of solutions, each depending on the arbitrary values of 'y', given by

$$x = \frac{1}{2} - \frac{3}{2}y$$

That is,

2 variables  1 equation  No unique solution

To summarize,

1 variable  1 equation  Unique solution

2 variables  1 equation  No unique solution

Ok, what about

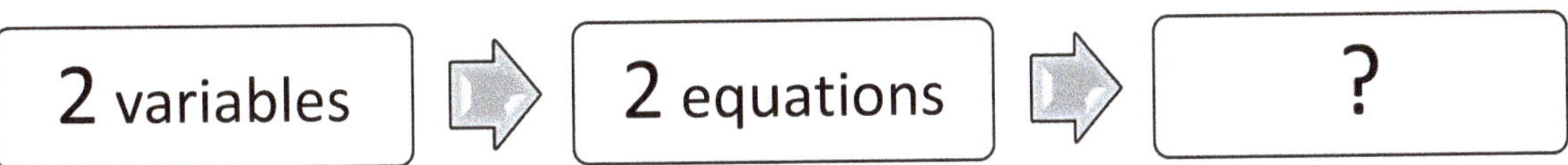

To find out, let's consider two linear equations having two variables. A set of two or more linear equations is called a linear system.

LINEAR SYSTEMS

Consider the system of two equations

$$\left.\begin{aligned} 2x + 3y &= 1 \\ 5x - 3y &= 34 \end{aligned}\right\}$$

Upon plotting the two equations,

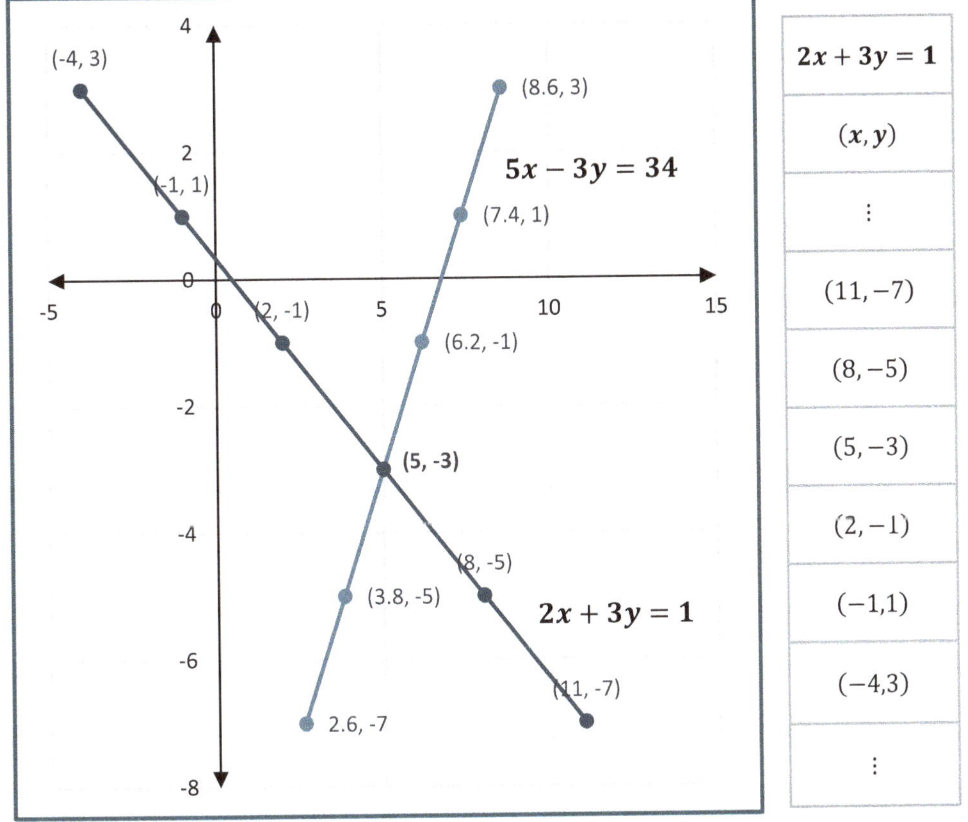

$2x + 3y = 1$
(x, y)
⋮
$(11, -7)$
$(8, -5)$
$(5, -3)$
$(2, -1)$
$(-1, 1)$
$(-4, 3)$
⋮

$5x - 3y = 34$
(x, y)
⋮
$(2.6, -7)$
$(3.8, -5)$
$(5, -3)$
$(6.2, -1)$
$(7.4, 1)$
$(8.6, 3)$
⋮

We find that they intersect each other at $(5, -3)$.

Ok, what's going on here?

To find out, let's substitute the values of 'x' and 'y' at the point of intersection. That is

$$x = 5$$
$$y = -3$$

back into the two equations.

$2x + 3y = 1$	$5x - 3y = 34$
$\Rightarrow 2(5) + 3(-3) = 1$	$\Rightarrow 5(5) - 3(-3) = 34$
or	or
$1 = 1$	$34 = 34$

Here, the values $x = 5$ and $y = -3$ respects the equality of not just one equation, but both of them simultaneously. Meaning, the set of values

$$\left.\begin{matrix} x = 5 \\ y = -3 \end{matrix}\right\}$$

solves both the equations

$$2x + 3y = 1$$
$$5x - 3y = 34$$

at the same time.

Thus, they are called the '**simultaneous solutions**' of the linear system. They are represented as the ordered pair $(5, -3)$ or as the set of equations

$$\left.\begin{matrix} x = 5 \\ y = -3 \end{matrix}\right\}$$

called the solution set.

In a nutshell, **the solution of a linear system is its point of intersection**. And, since two straight lines cannot intersect at more than one point, the solution $(5, -3)$ is a unique solution. That is, we conclude,

2 variables  2 equations  Unique solution

Following the pattern,

1 variable → 1 equation → Unique solution

2 variables → 1 equation → No unique solution

2 variables → 2 equations → Unique solution

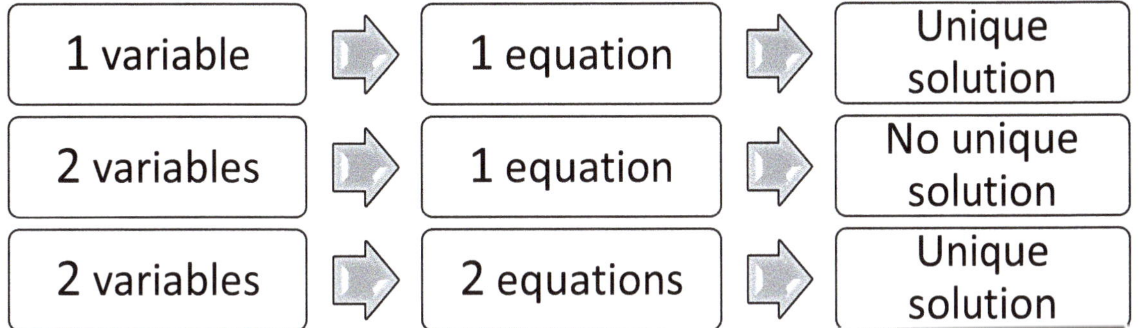

we naively deduce a rule of thumb, that a **linear system with a unique solution**, has

Number of variables = Number of equations

Problem 29: Which of the following systems can have unique solutions.

i. $$\left.\begin{aligned} 2x + 8y + z &= 0 \\ -5x + 2y - 9z &= 26 \\ 7y + 3z &= -12 \end{aligned}\right\}$$

ii. $$\left.\begin{aligned} 2x + 8y + z &= 0 \\ -5x + 2y - 9z &= 0 \end{aligned}\right\}$$

iii. $$\left.\begin{aligned} 6x + 4y &= 2 \\ 3x - 5y &= -34 \end{aligned}\right\}$$

Ans:

i. 3 variables → 3 equations → Unique solution

ii. 3 variables → 2 equations → No unique solution

iii. 2 variables → 2 equations → Unique solution

Note that this rule is not universal. It breaks down if the linear system is inconsistent. Which we shall see in the upcoming pages.

SPECIAL CASES OF LINEAR SYSTEMS

I- Linear systems with infinite solutions

Consider the system

$$\left.\begin{aligned} x + 2y &= 1 \\ 2x + 4y &= 2 \end{aligned}\right\}$$

Here, the second equation can be converted into the first as,

$$2(x + 2y) = 2$$

or

$$x + 2y = 1$$

Since both the equations are the same, the system becomes,

$$\left.\begin{aligned} x + 2y &= 1 \\ x + 2y &= 1 \end{aligned}\right\}$$

Thus, even though at first glance, the system seemed to have two equations with two variables, by algebraic manipulations, we find that it is actually one equation with two variables.

Thus, the system has infinite solutions given by,

$$x = 1 - 2y$$

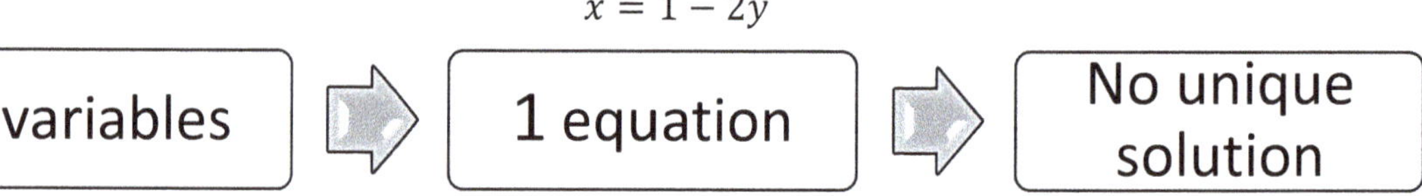

See the figure,

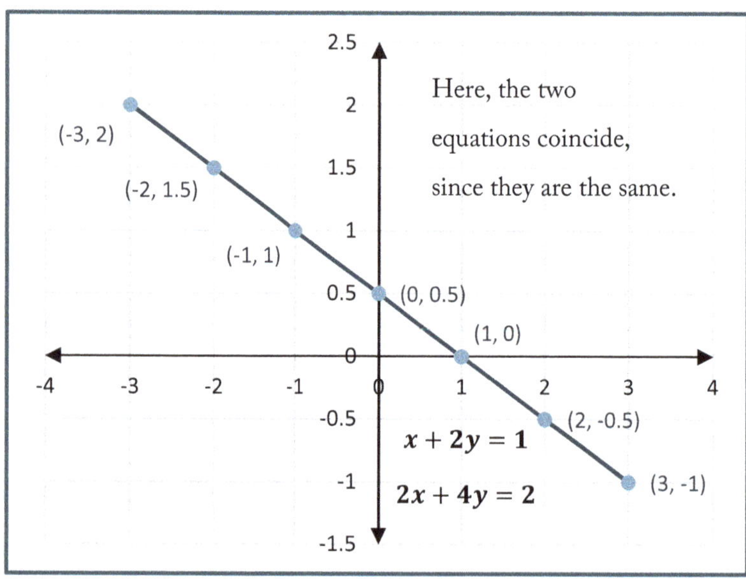

$\boldsymbol{x + 2y = 1}$
$\boldsymbol{(x, y)}$
$\vdots$
$(-3,2)$
$(-2,1.5)$
$(-1,1)$
$(0,0.5)$
$(1,0)$
$(2,-0.5)$
$(3,-1)$
$\vdots$

Another example is the system,

$$\left.\begin{aligned} 2x + 3y + 5z &= 7 \\ x + 2y + 3z &= 1 \\ 3x + 6y + 9z &= 3 \end{aligned}\right\}$$

where, the third equation is equivalent to the second one. That is

$$3(x + 2y + 3z) = 3$$

or

$$x + 2y + 3z = 1$$

Thus, the system becomes,

$$\left.\begin{aligned} 2x + 3y + 5z &= 7 \\ x + 2y + 3z &= 1 \\ x + 2y + 3z &= 1 \end{aligned}\right\}$$

with three variables and two equations. Therefore, has no unique solution.

2 equations

No unique solution

II- Inconsistent Linear systems

Consider the system,

$$\left.\begin{aligned} x + y &= 1 \\ x + y &= 0 \end{aligned}\right\}$$

Since there are two equations and two variables, it is expected to have a unique solution.

But, upon plotting,

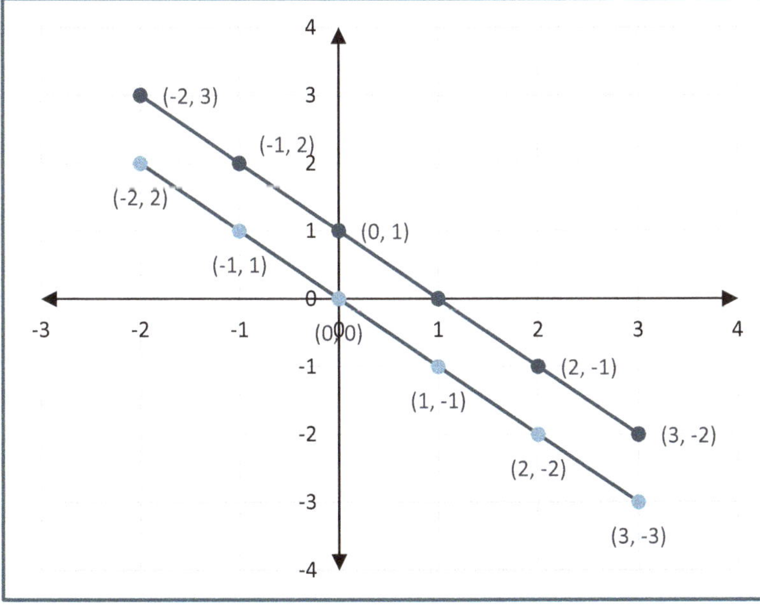

$x + y = 1$
(x, y)
$\vdots$
$(-2,3)$
$(-1,2)$
$(0,1)$
$(1,0)$
$(2,-1)$
$(3,-2)$
$\vdots$

$x + y = 0$
(x, y)
$\vdots$
$(-2,2)$
$(-1,1)$
$(0,0)$
$(1,-1)$
$(2,-2)$
$(3,-3)$
$\vdots$

we find that the equations do not intersect each other, nor do they coincide. This shows that the system has no solutions at all. Such systems are said to be **inconsistent**.

III- Homogenous Linear systems

A linear system is homogenous if all of its constant terms in the RHS are zero. Examples are,

$$\left.\begin{matrix} 2x+3y=0 \\ 5x-3y=0 \end{matrix}\right\} \; ; \; \left.\begin{matrix} x-5y=0 \\ 3x-15y=0 \end{matrix}\right\} \text{ etc.}$$

A homogenous system can have only two types of solutions.

Zero Solution

The zero solution is common to all homogeneous systems. It is obtained by assigning zero to each variable. That is,

$$\left.\begin{matrix} x=0 \\ y=0 \end{matrix}\right\}$$

Upon substituting the zero solution into the above examples,

$$\left.\begin{matrix} 2x+3y=0 \\ 5x-3y=0 \end{matrix}\right\}$$

becomes,

$$\left.\begin{matrix} 0=0 \\ 0=0 \end{matrix}\right\}$$

$$\left.\begin{matrix} x-5y=0 \\ 3x-15y=0 \end{matrix}\right\}$$

becomes,

$$\left.\begin{matrix} 0=0 \\ 0=0 \end{matrix}\right\}$$

This shows the universality of the zero solution for all homogenous systems.

Infinite Solutions

Other than the zero solution, some homogenous systems can have infinite solutions. This is possible if one of the equations is equivalent to another. For instance, the system

$$\left.\begin{matrix} 2x+3y=0 \\ 5x-3y=0 \end{matrix}\right\}$$

has only the zero solution, since the two equations are not equivalent, or in other words, one cannot be converted into another.

But in the system

$$\left.\begin{matrix} x-5y=0 \\ 3x-15y=0 \end{matrix}\right\}$$

the second equation can be converted into the first as,

$$3(x - 5y) = 0$$

or

$$x - 5y = 0$$

That is, the system is equivalent to the equation,

$$x - 5y = 0$$

having two variables with infinite solutions given by,

$$\left.\begin{aligned} x &= 5y \\ y &= y \end{aligned}\right\}$$

Ok let's recap.

- **A linear equation with one variable** can have a **unique** solution that can be represented as a **point** on the **number line.**
- **A linear equation with two variables** can have **infinite number of solutions-(x, y)**, forming a **straight line** in the X-Y coordinate system.
- A set of two or more linear equations is called a **linear system.**
- Linear systems come in two varieties: **non-homogeneous** (non-zero constant terms in the RHS) and **homogeneous** (zero constant terms in the RHS).
- For a non-homogenous system either:
  - The equations **intersect** each other, giving a **unique solution.**
  - The equations **coincide** each other, giving **infinite solutions** (this happens if **one of the equations can be converted into another**).
  - The equations **never meet,** giving no solution (such systems are said to be **inconsistent**).
- For a homogenous system either:
  - The system has the **zero solution,** obtained by setting the **variables equal to zero.**
  - The system has **infinite solutions,** when **one of the equations can be converted into another.**
- **Remember:** for both homogenous and non-homogenous systems, **if one of the equations can be converted into another, the number of variables will become greater than the number of equations,** since one equation is reduced into another.

This **results in additional variables in the solution whose arbitrary values gives infinite solutions.**

Example: $2x + 3y = 1$; $\left.\begin{matrix} x + 2y = 1 \\ 2x + 4y = 2 \end{matrix}\right\}$; $\left.\begin{matrix} x - 5y = 0 \\ 3x - 15y = 0 \end{matrix}\right\}$ etc.

To summarize,

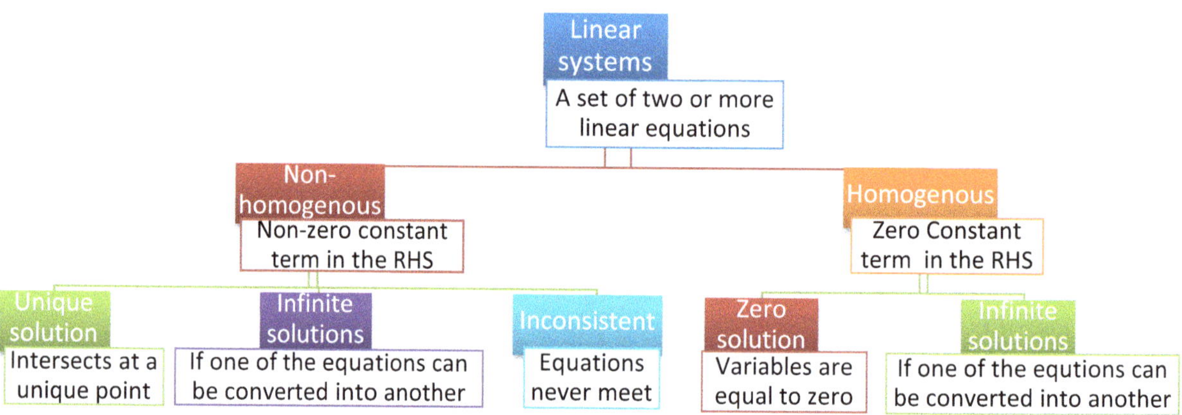

Linear equations form the basic foundation upon which concepts like matrices and determinants are built. So, take your time and study them thoroughly. If you are confident, continue to the next session. Otherwise, keep working until you are clear.

SESSION THREE

CONNECTING MATRIX AND LINEAR EQUATIONS

We have studied matrices; we have studied linear equations. Now it's time to combine them both. The connection is surprisingly simple.

Since the equality of two matrices is the equality of their corresponding elements, a linear system for instance,

$$\left.\begin{aligned} 2x + 3y &= 1 \\ 5x - 3y &= 34 \end{aligned}\right\}$$

is equivalent to the matrix equation,

$$\begin{pmatrix} 2x + 3y \\ 5x - 3y \end{pmatrix} = \begin{pmatrix} 1 \\ 34 \end{pmatrix}$$

And the matrix on the LHS

$$\begin{pmatrix} 2x + 3y \\ 5x - 3y \end{pmatrix}$$

is equivalent to the product

$$\begin{pmatrix} 2 & 3 \\ 5 & -3 \end{pmatrix}\begin{pmatrix} x \\ y \end{pmatrix}$$

That is,

$$\underset{2\times\boxed{2=2}\times 1}{\begin{pmatrix} 2 & 3 \\ 5 & -3 \end{pmatrix}\begin{pmatrix} x \\ y \end{pmatrix}} = \underset{2\times 1}{\begin{pmatrix} 2x + 3y \\ 5x - 3y \end{pmatrix}} = \underset{2\times 1}{\begin{pmatrix} 1 \\ 34 \end{pmatrix}}$$

where, the matrix

$$\begin{pmatrix} 2 & 3 \\ 5 & -3 \end{pmatrix}$$

is called the **coefficient matrix**, since it contains all the coefficients of the system.

Thus, a general linear system of two variables

$$\left.\begin{aligned} a_{11}x_1 + a_{12}x_2 &= b_1 \\ a_{21}x_1 + a_{22}x_2 &= b_2 \end{aligned}\right\}$$

can be written as the matrix equation

$$\begin{pmatrix} a_{11} & a_{12} \\ a_{21} & a_{22} \end{pmatrix}\begin{pmatrix} x_1 \\ x_2 \end{pmatrix} = \begin{pmatrix} b_1 \\ b_2 \end{pmatrix}$$

And a general linear system of three variables

$$\left.\begin{aligned} a_{11}x_1 + a_{12}x_2 + a_{13}x_3 &= b_1 \\ a_{21}x_1 + a_{22}x_2 + a_{23}x_3 &= b_2 \\ a_{31}x_1 + a_{32}x_2 + a_{33}x_3 &= b_3 \end{aligned}\right\}$$

can be written as the matrix equation,

$$\begin{pmatrix} a_{11} & a_{12} & a_{13} \\ a_{21} & a_{22} & a_{23} \\ a_{31} & a_{32} & a_{33} \end{pmatrix} \begin{pmatrix} x_1 \\ x_2 \\ x_3 \end{pmatrix} = \begin{pmatrix} b_1 \\ b_2 \\ b_3 \end{pmatrix}$$

This method is naturally extended to systems of any number of variables as,

$$\mathbf{AX} = \mathbf{B}$$

where,

$$\mathbf{A} = \begin{pmatrix} a_{11} & a_{12} & \cdots & a_{1n} \\ a_{21} & a_{22} & \cdots & a_{2n} \\ \vdots & \vdots & \ddots & \vdots \\ a_{n1} & a_{n2} & \cdots & a_{nn} \end{pmatrix} \; ; \; \mathbf{X} = \begin{pmatrix} x_1 \\ x_2 \\ \vdots \\ x_n \end{pmatrix} \text{ and } \mathbf{B} = \begin{pmatrix} b_1 \\ b_2 \\ \vdots \\ b_n \end{pmatrix}$$

Problem 30: Express the following linear systems in terms of matrices.

i. $\left.\begin{aligned} 2x + 5y &= 10 \\ 3x + 4y &= 24 \end{aligned}\right\}$

ii. $\left.\begin{aligned} x + y - z &= 1 \\ x + y + z &= 3 \\ x - y &= 4 \end{aligned}\right\}$

iii. $\left.\begin{aligned} x_1 &= 3 \\ 2x_1 + x_2 &= 4 \end{aligned}\right\}$

iv. $\left.\begin{aligned} x &= 1 \\ y &= 2 \\ z &= 3 \end{aligned}\right\}$

Ans:

i. $\begin{pmatrix} 2 & 5 \\ 3 & 4 \end{pmatrix} \begin{pmatrix} x \\ y \end{pmatrix} = \begin{pmatrix} 10 \\ 24 \end{pmatrix}$

ii. $\begin{pmatrix} 1 & 1 & -1 \\ 1 & 1 & 1 \\ 1 & -1 & 0 \end{pmatrix} \begin{pmatrix} x \\ y \\ z \end{pmatrix} = \begin{pmatrix} 1 \\ 3 \\ 4 \end{pmatrix}$

iii. $\begin{pmatrix} 1 & 0 \\ 2 & 1 \end{pmatrix} \begin{pmatrix} x_1 \\ x_2 \end{pmatrix} = \begin{pmatrix} 3 \\ 4 \end{pmatrix}$

iv. $\begin{pmatrix} 1 & 0 & 0 \\ 0 & 1 & 0 \\ 0 & 0 & 1 \end{pmatrix} \begin{pmatrix} x \\ y \\ z \end{pmatrix} = \begin{pmatrix} 1 \\ 2 \\ 3 \end{pmatrix}$

The separation of coefficients and variables into two distinct matrices makes the system more tidy and well organised. It also simplifies the methods by which the system can be solved, which is the topic of the next chapter.

But, is this connection between matrices and linear equations a coincidence?

Actually not!

In fact, the very idea of matrices is derived from linear systems.

Let's look at it this way.

Let's say we have a system,

$$\left.\begin{aligned} 2x + 3y &= 1 \\ 5x - 3y &= 34 \end{aligned}\right\}$$

For neatness, let's put the LHS and the RHS in two boxes

$$\begin{pmatrix} 2x + 3y \\ 5x - 3y \end{pmatrix} \text{ and } \begin{pmatrix} 1 \\ 34 \end{pmatrix}$$

Let's call them "**matrices**".

In order for the equality of the system to be valid, the corresponding elements of the boxes should be equal. That is,

$$\begin{pmatrix} 2x + 3y \\ 5x - 3y \end{pmatrix} = \begin{pmatrix} 1 \\ 34 \end{pmatrix} \text{ means } \begin{matrix} 2x + 3y = 1 \\ \text{and} \\ 5x - 3y = 34 \end{matrix}$$

This is why the equality of matrices is defined as the equality of the corresponding elements, so that it respects the equality of the linear system.

Now, the separation of coefficients and variables into two distinct matrices

$$\begin{pmatrix} 2 & 3 \\ 5 & -3 \end{pmatrix} \text{ and } \begin{pmatrix} x \\ y \end{pmatrix}$$

demands the definition of a new type of product, where the elements of the corresponding rows and columns are multiplied and added in order to obtain the elements of the LHS matrix. That is,

$$\begin{pmatrix} 2 & 3 \\ 5 & -3 \end{pmatrix}\begin{pmatrix} x \\ y \end{pmatrix} = \begin{pmatrix} (2)(x) + (3)(y) \\ (5)(x) + (-3)(y) \end{pmatrix} = \begin{pmatrix} 2x + 3y \\ 5x - 3y \end{pmatrix}$$

No other way of multiplication can result in the LHS matrix. That's why matrix multiplication is defined in this way.

Day Five

Session One

Solving a system using inverse

Limitations of the inverse method

Session Two

Gaussian elimination

Session Three

Rank

Inconsistent systems

Systems with infinite solutions

Homogenous systems

What you do everyday matters more than what you do once in a while.

- Gretchen Rubin

SESSION ONE

Take a few breaths and meditate on everything we have studied so far. If you have flash cards, go through them with patience and focus. Try to recall and reconnect with each and every topic.

Today, we will learn how matrix methods can be used to solve linear systems.

First, we present the inverse method.

SOLVING A LINEAR SYSTEM USING INVERSE

See the example.

Example 8: Solve the following linear system using the inverse method.

$$\left.\begin{aligned} 2x + 3y &= 1 \\ 5x - 3y &= 34 \end{aligned}\right\}$$

Solution: In matrix form the system becomes,

$$\begin{pmatrix} 2 & 3 \\ 5 & -3 \end{pmatrix}\begin{pmatrix} x \\ y \end{pmatrix} = \begin{pmatrix} 1 \\ 34 \end{pmatrix}$$

With a coefficient matrix,

$$A = \begin{pmatrix} 2 & 3 \\ 5 & -3 \end{pmatrix}$$

Now, find the inverse of the coefficient matrix using row operations. That is,

$$(A \mid I) = \left(\begin{array}{cc|cc} 2 & 3 & 1 & 0 \\ 5 & -3 & 0 & 1 \end{array}\right)$$

$R_1 \longrightarrow \frac{1}{2}R_1$

$$\left(\begin{array}{cc|cc} 1 & 3/2 & 1/2 & 0 \\ 5 & -3 & 0 & 1 \end{array}\right)$$

$R_2 \longrightarrow R_2 - 5R_1$

$$\left(\begin{array}{cc|cc} 1 & 3/2 & 1/2 & 0 \\ 0 & -21/2 & -5/2 & 1 \end{array}\right)$$

$R_2 \longrightarrow -\frac{2}{21}R_2$

$$\left(\begin{array}{cc|cc} 1 & ^3/_2 & ^1/_2 & 0 \\ 0 & 1 & ^5/_{21} & ^{-2}/_{21} \end{array}\right)$$

$R_1 \longrightarrow R_1 - \frac{3}{2}R_2$

$$\left(\begin{array}{cc|cc} 1 & 0 & ^3/_{21} & ^3/_{21} \\ 0 & 1 & ^5/_{21} & ^{-2}/_{21} \end{array}\right)$$

Thus

$$A^{-1} = \begin{pmatrix} ^3/_{21} & ^3/_{21} \\ ^5/_{21} & ^{-2}/_{21} \end{pmatrix}$$

Now, multiply the system by the inverse. That is,

$$\begin{pmatrix} ^3/_{21} & ^3/_{21} \\ ^5/_{21} & ^{-2}/_{21} \end{pmatrix}\begin{pmatrix} 2 & 3 \\ 5 & -3 \end{pmatrix}\begin{pmatrix} x \\ y \end{pmatrix} = \begin{pmatrix} ^3/_{21} & ^3/_{21} \\ ^5/_{21} & ^{-2}/_{21} \end{pmatrix}\begin{pmatrix} 1 \\ 34 \end{pmatrix}$$

Since,

$$A^{-1}A = I = \begin{pmatrix} 1 & 0 \\ 0 & 1 \end{pmatrix}$$

we get,

$$\begin{pmatrix} 1 & 0 \\ 0 & 1 \end{pmatrix}\begin{pmatrix} x \\ y \end{pmatrix} = \begin{pmatrix} 5 \\ -3 \end{pmatrix}$$

or,

$$\begin{pmatrix} x \\ y \end{pmatrix} = \begin{pmatrix} 5 \\ -3 \end{pmatrix}$$

Using the equality of the matrix, we get the required solution,

$$\left.\begin{array}{r} x = 5 \\ y = -3 \end{array}\right\}$$

(See, Day Four- Session Two)

One should always double check the solution by substituting back into the original equation.

Example 9: Solve the following linear system using inverse method.

$$\left.\begin{array}{c}2x_1 + x_2 - x_3 = 2\\ 2x_1 + x_3 = 3\\ x_1 - x_2 = 0\end{array}\right\}$$

Solution: In the matrix form, the system becomes,

$$\begin{pmatrix}2 & 1 & -1\\ 2 & 0 & 1\\ 1 & -1 & 0\end{pmatrix}\begin{pmatrix}x_1\\ x_2\\ x_3\end{pmatrix} = \begin{pmatrix}2\\ 3\\ 0\end{pmatrix}$$

With a coefficient matrix,

$$A = \begin{pmatrix}2 & 1 & -1\\ 2 & 0 & 1\\ 1 & -1 & 0\end{pmatrix}$$

whose inverse is found to be,

$$A^{-1} = \begin{pmatrix}1/5 & 1/5 & 1/5\\ 1/5 & 1/5 & -4/5\\ -2/5 & 3/5 & -2/5\end{pmatrix}$$

Multiplying the system by the inverse gives.

$$\begin{pmatrix}1 & 0 & 0\\ 0 & 1 & 0\\ 0 & 0 & 1\end{pmatrix}\begin{pmatrix}x_1\\ x_2\\ x_3\end{pmatrix} = \begin{pmatrix}1/5 & 1/5 & 1/5\\ 1/5 & 1/5 & -4/5\\ -2/5 & 3/5 & -2/5\end{pmatrix}\begin{pmatrix}2\\ 3\\ 0\end{pmatrix} = \begin{pmatrix}1\\ 1\\ 1\end{pmatrix}$$

That is,

$$\begin{pmatrix}x_1\\ x_2\\ x_3\end{pmatrix} = \begin{pmatrix}1\\ 1\\ 1\end{pmatrix}$$

or,

$$\left.\begin{array}{c}x_1 = 1\\ x_2 = 1\\ x_3 = 1\end{array}\right\}$$

Problem 31: Solve the following linear systems using inverse method.

i. $\left.\begin{aligned} 2x - y &= 1 \\ 3x + 2y &= 12 \end{aligned}\right\}$ ii. $\left.\begin{aligned} x + y &= -2 \\ x - 2y &= 7 \end{aligned}\right\}$ iii. $\left.\begin{aligned} x + 2y - z &= 1 \\ 2x + y + 4z &= 2 \\ 3x + 3y + 4z &= 1 \end{aligned}\right\}$

Ans:

i. $\left.\begin{aligned} x &= 2 \\ y &= 3 \end{aligned}\right\}$ ii. $\left.\begin{aligned} x &= 1 \\ y &= -3 \end{aligned}\right\}$ iii. $\left.\begin{aligned} x &= 7 \\ y &= -4 \\ z &= -2 \end{aligned}\right\}$

LIMITATIONS OF THE INVERSE METHOD

The inverse method fails if the system has infinite solutions. For instance, consider the linear system having infinite solutions,

$$\left.\begin{aligned} 2x - 3y &= 6 \\ 6x - 9y &= 18 \end{aligned}\right\}$$

In matrix form the system becomes,

$$\begin{pmatrix} 2 & -3 \\ 6 & -9 \end{pmatrix}\begin{pmatrix} x \\ y \end{pmatrix} = \begin{pmatrix} 6 \\ 18 \end{pmatrix}$$

With coefficient matrix,

$$\mathbf{A} = \begin{pmatrix} 2 & -3 \\ 6 & -9 \end{pmatrix}$$

Now, finding the inverse of the coefficient matrix,

$$(\mathbf{A} \mid \mathbf{I}) = \left(\begin{array}{cc|cc} 2 & -3 & 1 & 0 \\ 6 & -9 & 0 & 1 \end{array}\right)$$

$R_1 \longrightarrow \frac{1}{2} R_1$

$$\left(\begin{array}{cc|cc} 1 & -3/2 & 1/2 & 0 \\ 6 & -9 & 0 & 1 \end{array}\right)$$

$R_2 \longrightarrow R_2 - 6R_1$

$$\left(\begin{array}{cc|cc} 1 & -3/2 & 1/2 & 0 \\ 0 & 0 & -3 & 1 \end{array}\right)$$

The appearance of a zero row during the row transformation of the matrix **A** shows that **A** has no inverse.

In such a case, the inverse method is useless.

Therefore, a more general method known as the **Gaussian elimination** is preferred.

SESSION TWO

GAUSSIAN ELIMINATION

The Gaussian elimination involves the reduction of the coefficient matrix into a row-echelon form, in order to eliminate the variables of the linear system. It is named after the famous German mathematician Carl Friedrich Gauss (1777-1855).

See the example below,

Example 10: Solve the following linear system using Gaussian elimination.

$$\left.\begin{matrix} 2x + 3y = 1 \\ 5x - 3y = 34 \end{matrix}\right\}$$

Solution: In matrix form the system becomes,

$$\begin{pmatrix} 2 & 3 \\ 5 & -3 \end{pmatrix}\begin{pmatrix} x \\ y \end{pmatrix} = \begin{pmatrix} 1 \\ 34 \end{pmatrix}$$

Let,

$$A = \begin{pmatrix} 2 & 3 \\ 5 & -3 \end{pmatrix} \; ; \; X = \begin{pmatrix} x \\ y \end{pmatrix} \text{ and } B = \begin{pmatrix} 1 \\ 34 \end{pmatrix}$$

So that,

$$AX = B$$

The first step is to form a Block matrix by joining the coefficient matrix A with the RHS matrix B. That is,

$$(\mathbf{A} \mid \mathbf{B}) = \left(\begin{array}{cc|c} 2 & 3 & 1 \\ 5 & -3 & 34 \end{array}\right)$$

This matrix is called the Augmented matrix. An augmented matrix contains all the numbers in a linear system.

Now, use the elementary row operations on the entire augmented matrix to transform A into a row-echelon form. That is,

$$(A \mid B) = \left(\begin{array}{cc|c} 2 & 3 & 1 \\ 5 & -3 & 34 \end{array}\right)$$

$R_1 \longrightarrow \frac{1}{2}R_1$

$$\left(\begin{array}{cc|c} 1 & 3/2 & 1/2 \\ 5 & -3 & 34 \end{array}\right)$$

$R_2 \longrightarrow R_2 - 5R_1$

$$\left(\begin{array}{cc|c} 1 & 3/2 & 1/2 \\ 0 & -21/2 & 63/2 \end{array}\right)$$

$R_2 \longrightarrow -\frac{2}{21}R_2$

$$\left(\begin{array}{cc|c} 1 & 3/2 & 1/2 \\ 0 & 1 & -3 \end{array}\right)$$

Here, the matrix A is reduced into the row-echelon matrix

$$\begin{pmatrix} 1 & 3/2 \\ 0 & 1 \end{pmatrix}$$

And the RHS matrix is reduced into

$$\begin{pmatrix} 1/2 \\ -3 \end{pmatrix}$$

Now, replace the coefficient matrix and the RHS matrix in the original equation with these two matrices. That is,

$$\begin{pmatrix} 1 & 3/2 \\ 0 & 1 \end{pmatrix}\begin{pmatrix} x \\ y \end{pmatrix} = \begin{pmatrix} 1/2 \\ -3 \end{pmatrix}$$

Which gives,

$$\begin{pmatrix} x + \frac{3}{2}y \\ y \end{pmatrix} = \begin{pmatrix} 1/2 \\ -3 \end{pmatrix}$$

or,

$$x + \frac{3}{2}y = \frac{1}{2} \text{ and } y = -3$$

Here, in the second equation, note that the variable 'x' is eliminated and the value of 'y' is obtained as,

$$y = -3$$

This elimination was possible because of the reduction of A into a row-echelon form.

Now, to find 'x', substitute the value of 'y' into the first equation. That is,

$$x + \frac{3}{2}(-3) = \frac{1}{2}$$

giving,

$$x = 5$$

Thus, the required solution is

$$\left.\begin{matrix} x = 5 \\ y = -3 \end{matrix}\right\} \text{ or } (5, -3)$$

Example 11: Solve the following linear system using Gaussian elimination.

$$\left.\begin{matrix} 2x_1 + x_2 - x_3 = 2 \\ 2x_1 + x_3 = 3 \\ x_1 - x_2 = 0 \end{matrix}\right\}$$

Solution: In matrix form the system becomes,

$$\begin{pmatrix} 2 & 1 & -1 \\ 2 & 0 & 1 \\ 1 & -1 & 0 \end{pmatrix} \begin{pmatrix} x_1 \\ x_2 \\ x_3 \end{pmatrix} = \begin{pmatrix} 2 \\ 3 \\ 0 \end{pmatrix}$$

Here, the augmented matrix is

$$(A \mid B) = \left(\begin{array}{ccc|c} 2 & 1 & -1 & 2 \\ 2 & 0 & 1 & 3 \\ 1 & -1 & 0 & 0 \end{array}\right)$$

Employing row operations to obtain the row-echelon form of A:

$R_1 \longrightarrow \frac{1}{2}R_1$

$$\left(\begin{array}{ccc|c} 1 & 1/2 & -1/2 & 1 \\ 2 & 0 & 1 & 3 \\ 1 & -1 & 0 & 0 \end{array}\right)$$

$R_2 \longrightarrow R_2 - 2R_1$

$$\left(\begin{array}{ccc|c} 1 & 1/2 & -1/2 & 1 \\ 0 & -1 & 2 & 1 \\ 1 & -1 & 0 & 0 \end{array}\right)$$

$R_3 \longrightarrow R_3 - R_1$

$$\left(\begin{array}{ccc|c} 1 & 1/2 & -1/2 & 1 \\ 0 & -1 & 2 & 1 \\ 0 & -3/2 & 1/2 & -1 \end{array}\right)$$

$R_2 \longrightarrow -R_2$

$$\left(\begin{array}{ccc|c} 1 & 1/2 & -1/2 & 1 \\ 0 & 1 & -2 & -1 \\ 0 & -3/2 & 1/2 & -1 \end{array}\right)$$

$R_3 \longrightarrow R_3 + \frac{3}{2}R_2$

$$\left(\begin{array}{ccc|c} 1 & 1/2 & -1/2 & 1 \\ 0 & 1 & -2 & -1 \\ 0 & 0 & -5/2 & -5/2 \end{array}\right)$$

$R_3 \longrightarrow -\frac{2}{5}R_3$

$$\left(\begin{array}{ccc|c} 1 & 1/2 & -1/2 & 1 \\ 0 & 1 & -2 & -1 \\ 0 & 0 & 1 & 1 \end{array}\right)$$

Thus, the coefficient matrix is transformed into the row-echelon form,

$$\begin{pmatrix} 1 & 1/2 & -1/2 \\ 0 & 1 & -2 \\ 0 & 0 & 1 \end{pmatrix}$$

And the RHS matrix is transformed into,

$$\begin{pmatrix} 1 \\ -1 \\ 1 \end{pmatrix}$$

Therefore, the system becomes,

$$\begin{pmatrix} 1 & 1/2 & -1/2 \\ 0 & 1 & -2 \\ 0 & 0 & 1 \end{pmatrix}\begin{pmatrix} x_1 \\ x_2 \\ x_3 \end{pmatrix} = \begin{pmatrix} 1 \\ -1 \\ 1 \end{pmatrix}$$

That is,

$$\begin{pmatrix} x_1 + \frac{x_2}{2} - \frac{x_3}{2} \\ x_2 - 2x_3 \\ x_3 \end{pmatrix} = \begin{pmatrix} 1 \\ -1 \\ 1 \end{pmatrix}$$

which gives,

$$x_1 + \frac{x_2}{2} - \frac{x_3}{2} = 1$$
$$x_2 - 2x_3 = -1$$
$$x_3 = 1$$

Substituting the value of 'x_3' into the second equation gives,

$$x_2 - 2(1) = -1$$
or
$$x_2 = 1$$

Now, substituting the values of both 'x_3' and 'x_2' into the first equation gives,

$$x_1 + \frac{(1)}{2} - \frac{(1)}{2} = 1$$
or
$$x_1 = 1$$

thus, the solution is,

$$\left.\begin{matrix} x_1 = 1 \\ x_2 = 1 \\ x_3 = 1 \end{matrix}\right\} \text{ or } (1, 1, 1)$$

Problem 32: Solve the following linear systems using gaussian elimination.

i. $\left.\begin{aligned} 2x - y = 1 \\ 3x + 2y = 12 \end{aligned}\right\}$ ii. $\left.\begin{aligned} x + y = -2 \\ x - 2y = 7 \end{aligned}\right\}$ iii. $\left.\begin{aligned} x + 2y - z = 1 \\ 2x + y + 4z = 2 \\ 3x + 3y + 4z = 1 \end{aligned}\right\}$

Ans:

i. $\left.\begin{aligned} x = 2 \\ y = 3 \end{aligned}\right\}$ ii. $\left.\begin{aligned} x = 1 \\ y = -3 \end{aligned}\right\}$ iii. $\left.\begin{aligned} x = 7 \\ y = -4 \\ z = -2 \end{aligned}\right\}$

All the examples and problems in this session had unique solutions. In the next session, we will extend the gaussian elimination to systems having no solution. For that, we need to understand the concept of rank. Before that, take a break, meditate for a while, refresh and then come back.

SESSION THREE

RANK

The **rank** of a matrix is the **number of non-zero rows in its row-echelon form.** See the example.

Example 12: Find the rank of the matrix

$$\begin{pmatrix} 1 & 2 & 3 \\ 4 & 5 & 6 \\ 7 & 8 & 9 \end{pmatrix}$$

Solution: Rank is the number of non-zero rows in the row-echelon form.

The matrix

$$\begin{pmatrix} 1 & 2 & 3 \\ 4 & 5 & 6 \\ 7 & 8 & 9 \end{pmatrix}$$

has the row-echelon form

$$\begin{pmatrix} 1 & 2 & 3 \\ 0 & 1 & 2 \\ 0 & 0 & 0 \end{pmatrix} \text{ (see problem 25)}$$

with one zero-row and two non-zero rows.

$$\text{therefore the rank} = 2$$

Problem 33: Find the ranks of the following matrices.

i. $\begin{pmatrix} 7 & 2 & 11 \\ 6 & 3 & -3 \\ 5 & 0 & 0 \end{pmatrix}$ ii. $\begin{pmatrix} 2 & 3 \\ 6 & 9 \end{pmatrix}$

Ans:

i. **3** ii. **1**

A linear system is **consistent** if the **rank of the coefficient matrix is equal to the rank of the augmented matrix**. That is, the system

$$\mathbf{AX} = \mathbf{B}$$

is consistent, if

$$\text{rank of } \mathbf{A} = \text{rank of } (\mathbf{A} \mid \mathbf{B})$$

In **example: 10**, the rank of the coefficient matrix is found from the row-echelon form

$$\begin{pmatrix} 1 & 3/2 \\ 0 & 1 \end{pmatrix}$$

as 2. And that of the augmented matrix is found from the row-echelon form

$$\left(\begin{array}{cc|c} 1 & 3/2 & 1/2 \\ 0 & 1 & -3 \end{array}\right)$$

as also 2.

Since they are equal,

$$\text{rank of } \mathbf{A} = \text{rank of } (\mathbf{A} \mid \mathbf{B}) = 2$$

The system is consistent and has a solution,

$$\left.\begin{array}{r} x = 5 \\ y = -3 \end{array}\right\} \text{ or } (5, -3)$$

Problem 34: Which of the following linear systems are consistent.

i. $\left.\begin{array}{r} x + y = 1 \\ x + 2y = 5 \end{array}\right\}$ ii. $\left.\begin{array}{r} 5x - y = -2 \\ 10x - 2y = 12 \end{array}\right\}$ iii. $\left.\begin{array}{r} 2x_1 + x_2 - x_3 = 2 \\ 2x_1 + x_3 = 3 \\ x_1 - x_2 = 0 \end{array}\right\}$

Ans:

i. **and** iii.

INCONSISTENT SYSTEMS

A linear system is said to be **inconsistent** if it is **unsolvable** (see Day Four – Session Two).

In terms of matrices,

A linear system is **inconsistent** if the **rank of the coefficient matrix is not equal to the rank of the augmented matrix**. That is, the system

$$\mathbf{AX} = \mathbf{B}$$

is inconsistent, if

$$\text{rank of } \mathbf{A} \neq \text{rank of } (\mathbf{A} \mid \mathbf{B})$$

For instance, consider the inconsistent system,

$$\left.\begin{array}{l} x + y = 1 \\ x + y = 0 \end{array}\right\} \text{ (See Day Four-Session Two)}$$

In matrix form, the system becomes,

$$\begin{pmatrix} 1 & 1 \\ 1 & 1 \end{pmatrix} \begin{pmatrix} x \\ y \end{pmatrix} = \begin{pmatrix} 1 \\ 0 \end{pmatrix}$$

with an Augmented matrix,

$$(\mathbf{A} \mid \mathbf{B}) = \left(\begin{array}{cc|c} 1 & 1 & 1 \\ 1 & 1 & 0 \end{array}\right)$$

We transform this matrix into a row-echelon form,

$$\mathrm{R}_2 \longrightarrow \mathrm{R}_2 - \mathrm{R}_1$$

$$\left(\begin{array}{cc|c} 1 & 1 & 1 \\ 0 & 0 & -1 \end{array}\right)$$

$$\mathrm{R}_2 \longrightarrow -\mathrm{R}_2$$

$$\left(\begin{array}{cc|c} 1 & 1 & 1 \\ 0 & 0 & 1 \end{array}\right)$$

Now, the rank of the coefficient matrix is found from the row-echelon form

$$\begin{pmatrix} 1 & 1 \\ 0 & 0 \end{pmatrix}$$

as 1. And that of the augmented matrix is found from the row-echelon form

$$\left(\begin{array}{cc|c} 1 & 1 & 1 \\ 0 & 0 & 1 \end{array}\right)$$

as 2.

Since they are different,

$$1 \neq 2$$

The system is inconsistent and therefore unsolvable.

This is verified by substituting the matrices in the row-echelon form back into the system as,

$$\begin{pmatrix} 1 & 1 \\ 0 & 0 \end{pmatrix} \begin{pmatrix} x \\ y \end{pmatrix} = \begin{pmatrix} 1 \\ 1 \end{pmatrix}$$

or,

$$\begin{pmatrix} x + y \\ 0 \end{pmatrix} = \begin{pmatrix} 1 \\ 1 \end{pmatrix}$$

giving,

$$\left.\begin{aligned} x + y &= 1 \\ 0 &= 1 \end{aligned}\right\}$$

The absurd result in the second equation shows the inconsistency of the system.

Problem 35: Show that the system

$$\left.\begin{aligned} 2x + 3y &= 5 \\ 4x + 6y &= 6 \end{aligned}\right\}$$

is inconsistent.

SYSTEMS WITH INFINITE SOLUTIONS

See the following example,

Example 13: Solve the following system

$$\left.\begin{aligned} x + y + z &= -1 \\ 2x + 3y + 4z &= 4 \\ 2x + 2y + 2z &= -2 \end{aligned}\right\}$$

Solution: In matrix form the system becomes,

$$\begin{pmatrix} 1 & 1 & 1 \\ 2 & 3 & 4 \\ 2 & 2 & 2 \end{pmatrix} \begin{pmatrix} x \\ y \\ z \end{pmatrix} = \begin{pmatrix} -1 \\ 4 \\ -2 \end{pmatrix}$$

With an augmented matrix,

$$(\mathrm{A} \mid \mathrm{B}) = \left(\begin{array}{ccc|c} 1 & 1 & 1 & -1 \\ 2 & 3 & 4 & 4 \\ 2 & 2 & 2 & -2 \end{array}\right)$$

We transform this matrix into a row-echelon form.

$R_2 \longrightarrow R_2 - 2R_1$

$$\left(\begin{array}{ccc|c} 1 & 1 & 1 & -1 \\ 0 & 1 & 2 & 6 \\ 2 & 2 & 2 & -2 \end{array}\right)$$

$R_3 \longrightarrow R_3 - 2R_1$

$$\left(\begin{array}{ccc|c} 1 & 1 & 1 & -1 \\ 0 & 1 & 2 & 6 \\ 0 & 0 & 0 & 0 \end{array}\right)$$

Here,

$$\text{rank of A} = \text{rank of (A | B)} = 2$$

showing that the system is consistent.

Thus, the original system becomes,

$$\begin{pmatrix} 1 & 1 & 1 \\ 0 & 1 & 2 \\ 0 & 0 & 0 \end{pmatrix} \begin{pmatrix} x \\ y \\ z \end{pmatrix} = \begin{pmatrix} -1 \\ 6 \\ 0 \end{pmatrix}$$

or,

$$\begin{pmatrix} x + y + z \\ y + 2z \\ 0 \end{pmatrix} = \begin{pmatrix} -1 \\ 6 \\ 0 \end{pmatrix}$$

Giving,

$$\left.\begin{array}{c} x + y + z = -1 \\ y + 2z = 6 \\ 0 = 0 \end{array}\right\}$$

Solving the first equation for '*x*' and the second equation for '*y*' gives,

$$\left.\begin{array}{c} x = -y - z - 1 \\ y = 6 - 2z \end{array}\right\}$$

Notice that in the second equation, the value of 'y' is determined by 'z' alone.

If we substitute the value of '*y*' into the first equations, we get,

$$\left.\begin{array}{c} x = z - 7 \\ y = 6 - 2z \end{array}\right\}$$

Since, the variable 'z' appears only on the RHS, it is arbitrary, meaning it can have any possible value. And for each value of 'z' there is a unique 'x' and 'y', given by

$$\left.\begin{array}{l} x = z - 7 \\ y = 6 - 2z \end{array}\right\}$$

Thus, the solution becomes,

$$\left.\begin{array}{r} x = z - 7 \\ y = 6 - 2z \\ z = z \end{array}\right\} \text{ or } (z - 7, 6 - 2z, z)$$

Problem 36: Solve the following systems using gaussian elimination

i. $\left.\begin{array}{r} x + 2y = 1 \\ 2x + 4y = 2 \end{array}\right\}$

ii. $\left.\begin{array}{r} 2x + 3y + 5z = 7 \\ x + 2y + 3z = 1 \\ 3x + 6y + 9z = 3 \end{array}\right\}$

Ans:

i. $\left.\begin{array}{r} x = 1 - 2y \\ y = y \end{array}\right\}$

ii. $\left.\begin{array}{r} x = 11 - z \\ y = -5 - z \\ z = z \end{array}\right\}$

HOMOGENOUS LINEAR SYSTEMS

A homogenous linear system can have two types of solutions (See Day Four-Session Two)

TYPE-I: Zero Solution

See the example,

Example 14: Solve the following homogenous system.

$$\left.\begin{array}{r} 2x + 3y = 0 \\ 5x - 3y = 0 \end{array}\right\}$$

Solution: In the matrix form the system becomes,

$$\begin{pmatrix} 2 & 3 \\ 5 & -3 \end{pmatrix} \begin{pmatrix} x \\ y \end{pmatrix} = \begin{pmatrix} 0 \\ 0 \end{pmatrix}$$

with an augmented matrix

$$(A \mid B) = \left(\begin{array}{cc|c} 2 & 3 & 0 \\ 5 & -3 & 0 \end{array}\right)$$

Transforming the augmented matrix into a row-echelon form

$R_1 \longrightarrow \frac{1}{2}R_1$

$$\left(\begin{array}{cc|c} 1 & 3/2 & 0 \\ 5 & -3 & 0 \end{array}\right)$$

$R_2 \longrightarrow R_2 - 5R_1$

$$\left(\begin{array}{cc|c} 1 & 3/2 & 0 \\ 0 & -21/2 & 0 \end{array}\right)$$

$R_2 \longrightarrow -\frac{2}{21}R_2$

$$\left(\begin{array}{cc|c} 1 & 3/2 & 0 \\ 0 & 1 & 0 \end{array}\right)$$

giving

$$\left.\begin{array}{r} x + \frac{3}{2}y = 0 \\ y = 0 \end{array}\right\}$$

Substituting the value of 'y' into the first equation gives the solution,

$$\left.\begin{array}{r} x = 0 \\ y = 0 \end{array}\right\}$$

TYPE-II: Infinite Solutions

See the following example

Example 15: Solve the following homogenous system.

$$\left.\begin{array}{r} x - 5y = 0 \\ 3x - 15y = 0 \end{array}\right\}$$

Solution: In the matrix form the system becomes,

$$\begin{pmatrix} 1 & -5 \\ 3 & -15 \end{pmatrix}\begin{pmatrix} x \\ y \end{pmatrix} = \begin{pmatrix} 0 \\ 0 \end{pmatrix}$$

with an augmented matrix

$$(A \mid B) = \left(\begin{array}{cc|c} 1 & -5 & 0 \\ 3 & -15 & 0 \end{array}\right)$$

Transforming the augmented matrix into a row-echelon form

$\mathbf{R_2 \longrightarrow R_2 - 3R_1}$

$$\begin{pmatrix} 1 & -5 & | & 0 \\ 0 & 0 & | & 0 \end{pmatrix}$$

giving

$$\left.\begin{matrix} x - 5y = 0 \\ 0 = 0 \end{matrix}\right\} \text{ or } \left.\begin{matrix} x = 5y \\ 0 = 0 \end{matrix}\right\}$$

Since '*y*' is arbitrary, there are infinite solutions given by,

$$\left.\begin{matrix} x = 5y \\ y = y \end{matrix}\right\} \text{ or } (5y, y)$$

Problem 37: Solve the following systems using gaussian elimination

i. $$\left.\begin{matrix} 2x + 3y + 5z = 0 \\ 5x + 2y = z \\ 3x + 4y + 7z = 0 \end{matrix}\right\}$$

ii. $$\left.\begin{matrix} 6x + 4y - 2z = 0 \\ 2x + 3y + 5z = 0 \\ 3x + 2y - z = 0 \end{matrix}\right\}$$

Ans:

i. $$\left.\begin{matrix} x = 0 \\ y = 0 \\ z = 0 \end{matrix}\right\}$$

ii. $$\left.\begin{matrix} x = \frac{13z}{5} \\ y = -\frac{17z}{5} \\ z = z \end{matrix}\right\}$$

An effective way to improve our memory is by employing the '**spaced repetition**' strategy. The idea is to recall the topics in repeated sessions with enough time gaps, so that our brain is forced to understand its importance.

Also try to recall the previous session before recalling the current, so that our brain recognises where to place each of the topics in our memory.

Day Six

Session One
The Determinant

Session Two
Minors and Cofactors

Session Three
Determinants continued.

A shortcut for 3×3 matrices

Do not worry about your difficulties in mathematics, mine are still greater.

- Albert Einstein

SESSION ONE

THE DETEMINANT

The matrix methods of the previous sessions are immensely useful if the number of variables is large. But for systems of two variables, algebraic methods are also justified, since they are easy.

So, consider our old pal,

$$\left.\begin{array}{l}2x+3y=1\\5x-3y=34\end{array}\right\}$$

Find the value of 'y' from the first equation and substitute it into the second. That is, from

$$2x+3y=1$$

find,

$$y=\frac{1-2x}{3}$$

Substituting it into the second equation gives,

$$5x-3\left(\frac{1-2x}{3}\right)=34$$

after some algebra, we get,

$$(3)(5)x-(2)(-3)x-3=102$$

giving,

$$x=\frac{-105}{[(2)(-3)-(3)(5)]}=\frac{-105}{-21}=5$$

Similarly, for 'y' we get,

$$y=\frac{63}{[(2)(-3)-(3)(5)]}=\frac{63}{-21}=-3$$

Here, the two solutions,

$$x=\frac{-105}{[(2)(-3)-(3)(5)]}=5 \text{ and } y=\frac{63}{[(2)(-3)-(3)(5)]}=-3$$

have the same denominator

$$[(2)(-3)-(3)(5)]=-21$$

If the value of the denominator changes, both the solutions change simultaneously, since it appears on both.

From the coefficient matrix

$$\begin{pmatrix} 2 & 3 \\ 5 & -3 \end{pmatrix}$$

This denominator can be obtained by

i. **First:** Multiplying the prime diagonal elements,

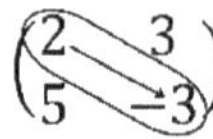

to get,

$$(2)(-3)$$

ii. **Second:** Multiplying the off-diagonal elements,

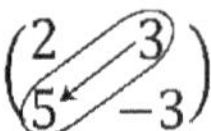

to get,

$$(3)(5)$$

iii. **Third:** Subtracting the later form the former, that is

$$(2)(-3) - (3)(5) = -21$$

This denominator plays a significant role in determining the nature of solutions. To understand this let's consider the inconsistent system,

$$\left.\begin{matrix} x + y = 1 \\ x + y = 0 \end{matrix}\right\}$$

Since the system is inconsistent, no value of 'x' and 'y' can satisfy these equations.

It is surprising to see that the denominator of the solutions reveals this fact.

To see that,

Let's solve the system algebraically,

From the first equation,

$$x + y = 1$$

find,

$$y = 1 - x$$

Substituting into the second gives,

$$x + 1 - x = 0$$

That is,

$$x[1 - 1] = 1$$

or,

$$x = \frac{1}{[1-1]} = \frac{1}{0}$$

Similarly, for 'y' we get,

$$y = \frac{1}{[1-1]} = \frac{1}{0}$$

Here, both the solutions have the same denominator- 'zero'.

Since, division by zero is not defined in mathematics (See page 94), no 'x' and 'y' can satisfy the relations,

$$x = \frac{1}{[1-1]} \text{ and } y = \frac{1}{[1-1]}$$

Hence, the system is unsolvable.

This result is a direct consequence of the value of the denominator.

Since this denominator **determines** whether a system is solvable or unsolvable, it is called the **'determinant'**.

The determinant is obtained from the coefficient matrix by subtracting the product of the off-diagonal elements from the product of the prime diagonal elements.

That is, from

$$\begin{pmatrix} 1 & 1 \\ 1 & 1 \end{pmatrix}$$

we obtain,

$$(1)(1) - (1)(1) = 0$$

Notice that the augmented matrix of the system

$$\left(\begin{array}{cc|c} 1 & 1 & 1 \\ 1 & 1 & 0 \end{array}\right)$$

do not have a determinant, since it doesn't have a diagonal.

Only square matrices can have determinants

The determinant of a matrix **A** is denoted by $\mathbf{det}(\mathbf{A})$ or $|\mathbf{A}|$.

Problem 38: Find the determinants of the following matrices.

i. $A = \begin{pmatrix} 1 & 2 \\ 3 & 4 \end{pmatrix}$ ii. $B = \begin{pmatrix} 1 & -1 \\ 0 & 1 \end{pmatrix}$ iii. $C = \begin{pmatrix} 1 & \sqrt{2} \\ \sqrt{2} & 1 \end{pmatrix}$

Ans:

i. $|A| = -2$ ii. $|B| = 1$ iii. $|C| = -1$

DIVISION BY ZERO

Division by zero is not defined in standard mathematics. To demonstrate this, let's divide the number 2 by zero.

$$\frac{2}{0}$$

Assume that this division is possible and an answer exists.

Let it be equal to 'x'. i.e.

$$\frac{2}{0} = x$$

Which gives,

$$2 = x(0)$$

Meaning, x times 0 equals 2.

No x can satisfy this equation since multiplication by zero always yields zero.

This shows that the number x cannot exist and the division $\frac{2}{0} = x$ is meaningless and therefore cannot be defined.

SESSION TWO

MINORS AND COFACTORS

The determinant of a 2×2 matrix,

$$\mathbf{A} = \begin{pmatrix} a_{11} & a_{12} \\ a_{21} & a_{22} \end{pmatrix}$$

is given by,

$$|\mathrm{A}| = \begin{vmatrix} a_{11} & a_{12} \\ a_{21} & a_{22} \end{vmatrix} = (a_{11})(a_{22}) - (a_{12})(a_{21})$$

To define the determinant of a 3×3 matrix, we need to define **minors** and **cofactors**.

For that, consider a 3×3 matrix,

$$\mathbf{A} = \begin{pmatrix} a_{11} & a_{12} & a_{13} \\ a_{21} & a_{22} & a_{23} \\ a_{31} & a_{32} & a_{33} \end{pmatrix}$$

The minor M_{11} is the determinant of the matrix obtained by deleting the row and column containing the element a_{11}, which is the first row-first column. That is,

$$\begin{pmatrix} a_{11} & a_{12} & a_{13} \\ a_{21} & a_{22} & a_{23} \\ a_{31} & a_{32} & a_{33} \end{pmatrix}$$

Delete this row

Delete this column

to obtain the matrix,

$$\begin{pmatrix} a_{22} & a_{23} \\ a_{32} & a_{33} \end{pmatrix}$$

Therefore,

$$\mathrm{M}_{11} = \begin{vmatrix} a_{22} & a_{23} \\ a_{32} & a_{33} \end{vmatrix} = (a_{22})(a_{33}) - (a_{23})(a_{32})$$

Similarly, M_{12} is the determinant of the matrix obtained by deleting the first row and second column. That is,

$$\begin{pmatrix} a_{11} & a_{12} & a_{13} \\ a_{21} & a_{22} & a_{23} \\ a_{31} & a_{32} & a_{33} \end{pmatrix}$$

Delete this row

Delete this column

to obtain the matrix,

$$\begin{pmatrix} a_{21} & a_{23} \\ a_{31} & a_{33} \end{pmatrix}$$

Therefore,

$$M_{12} = \begin{vmatrix} a_{21} & a_{23} \\ a_{31} & a_{33} \end{vmatrix} = (a_{21})(a_{33}) - (a_{23})(a_{31})$$

In general, the minor M_{ij} is the **determinant of the matrix obtained by deleting the ith row and j^{th} column** of a square matrix.

Problem 39: If

$$\mathbf{A} = \begin{pmatrix} 1 & 2 & 1 \\ 6 & -1 & 0 \\ -1 & -2 & -1 \end{pmatrix}$$

Find:

i. M_{11} **ii.** M_{12} **iii.** M_{33} **iv.** M_{23} **v.** M_{31}

Ans:

i. $M_{11} = 1$ **ii.** $M_{12} = -6$ **iii.** $M_{33} = -13$

iv. $M_{23} = 0$ **v.** $M_{31} = 1$

The **cofactor A_{ij}** corresponding to the minor M_{ij} of a square matrix is defined as,

$$A_{ij} = (-1)^{i+j} M_{ij}$$

Where,

$$(-1)^{i+j} = (-1)(-1)(-1) \ldots (i + j \text{ times})(-1)$$

For instance,

$$(-1)^2 = (-1)(-1) = 1$$

or,

$$(-1)^3 = (-1)(-1)(-1) = -1, \text{and so on} \ldots$$

See the example,

Example 16: Find the cofactors A_{11} and A_{23} of the matrix.

$$\begin{pmatrix} 2 & 3 & 1 \\ 0 & 2 & 1 \\ 3 & 4 & 8 \end{pmatrix}$$

Solution: To find A_{11}, first we have to find M_{11}.

$$M_{11} = \begin{vmatrix} 2 & 1 \\ 4 & 8 \end{vmatrix} = 12$$

Using M_{11} we get,

$$A_{11} = (-1)^{1+1}M_{11} = (-1)^2(12) = 12$$

Similarly,

$$M_{23} = \begin{vmatrix} 2 & 3 \\ 3 & 4 \end{vmatrix} = -1$$

Therefore,

$$A_{23} = (-1)^{2+3}M_{23} = (-1)(-1) = 1$$

Problem 40: Find the cofactors $A_{12}, A_{22}, A_{33}, A_{31}$ of the matrix

$$\begin{pmatrix} 1 & 2 & 4 \\ 6 & 7 & 5 \\ -1 & 5 & 1 \end{pmatrix}$$

Ans:

i. $A_{12} = -11$ ii. $A_{22} = 5$ iii. $A_{33} = -5$ iv. $A_{31} = -18$

Now at this point, it is good to take a pause and reflect on your progress. Recalling the topics is as important as learning them. So, recall and reconnect with each of them. Continue to the next session, only if you are clear.

In the next session, we will learn how to calculate the determinant of a general $n \times n$ matrix.

SESSION THREE

DETERMINANTS CONTINUED

Knowing how to calculate minors and cofactors, we can now define the determinant of a general $n \times n$ matrix.

Consider the simple case of a 3×3 matrix,

$$A = \begin{pmatrix} a_{11} & a_{12} & a_{13} \\ a_{21} & a_{22} & a_{23} \\ a_{31} & a_{32} & a_{33} \end{pmatrix}$$

i. **First:** Choose any row.

   In this example, let's choose the first row. That is,

$$(a_{11} \quad a_{12} \quad a_{13})$$

ii. **Second:** Find the cofactors corresponding to each element of the chosen row.

   For the first row, find

$$A_{11}, A_{12} \text{ and } A_{13}$$

iii. **Third:** Find the products of the elements of the chosen row and the corresponding cofactors.

   That is, find

$$a_{11}A_{11}, a_{12}A_{12} \text{ and } a_{13}A_{13}$$

iv. **Fourth:** Add them together to obtain the determinant.

   That is,

$$|A| = \begin{vmatrix} a_{11} & a_{12} & a_{13} \\ a_{21} & a_{22} & a_{23} \\ a_{31} & a_{32} & a_{33} \end{vmatrix} = a_{11}A_{11} + a_{12}A_{12} + a_{13}A_{13}$$

This method is known as the Laplacian expansion, named after the famous French mathematician Pierre Simon Laplace (1749-1827).

In the above example, the expansion was done along the first row. In fact, it can be done along any row of choice. **Always chose the row with the largest number of zeroes**, since it significantly reduces the calculations.

Example 17: Find the determinant of the matrix

$$A = \begin{pmatrix} 2 & 3 & -1 \\ 0 & 2 & 1 \\ 3 & 4 & 8 \end{pmatrix}$$

Solution: Let's expand along the first row, that is along

$$(2 \quad 3 \quad -1)$$

Where,

$$a_{11} = 2 \quad a_{12} = 3 \quad a_{13} = -1$$

Now,

$$M_{11} = \begin{vmatrix} 2 & 1 \\ 4 & 8 \end{vmatrix} = 12$$

$$M_{12} = \begin{vmatrix} 0 & 1 \\ 3 & 8 \end{vmatrix} = -3$$

$$M_{13} = \begin{vmatrix} 0 & 2 \\ 3 & 4 \end{vmatrix} = -6$$

From these minors, we get the cofactors,

$$A_{11} = (-1)^{1+1}M_{11} = (1)(12) = 12$$

$$A_{12} = (-1)^{1+2}M_{12} = (-1)(-3) = 3$$

$$A_{13} = (-1)^{1+3}M_{13} = (1)(-6) = -6$$

Therefore, we obtain the determinant as,

$$|A| = \begin{vmatrix} 2 & 3 & -1 \\ 0 & 2 & 1 \\ 3 & 4 & 8 \end{vmatrix} = a_{11}A_{11} + a_{12}A_{12} + a_{13}A_{13}$$

$$= (2)(12) + (3)(3) + (-1)(-6)$$

$$= 39$$

This method can be extended to square matrices of all orders.

See the example,

Example 18: Find the determinant of the 4×4 matrix

$$A = \begin{pmatrix} 1 & 4 & 2 & 3 \\ 7 & 3 & -5 & 2 \\ 3 & -5 & 0 & 0 \\ 4 & 2 & 8 & 1 \end{pmatrix}$$

Solution: Let's expand along the third row, since it contains the largest number of zeroes.

We have,

$$\begin{aligned} |A| &= a_{31}A_{31} + a_{32}A_{32} + a_{33}A_{33} + a_{34}A_{34} \\ &= (3)A_{31} + (-5)A_{32} + (0)A_{33} + (0)A_{34} \\ &= (3)A_{31} + (-5)A_{32} \end{aligned}$$

Now, M_{31} is the determinant,

$$\begin{vmatrix} 4 & 2 & 3 \\ 3 & -5 & 2 \\ 2 & 8 & 1 \end{vmatrix} = 20$$

Therefore,

$$A_{31} = (-1)^{3+1}M_{31} = (1)(20) = 20$$

Now, M_{32} is the determinant,

$$\begin{vmatrix} 1 & 2 & 3 \\ 7 & -5 & 2 \\ 4 & 8 & 1 \end{vmatrix} = 209$$

Therefore,

$$A_{32} = (-1)^{3+2}M_{32} = (-1)(209) = -209$$

Thus, we get,

$$\begin{aligned} |A| &= (3)(20) + (-5)(-209) \\ &= 1105 \end{aligned}$$

Problem 41: **Find the determinants of the following matrices,**

i. $\begin{pmatrix} 1 & 2 & 3 \\ 4 & 5 & 6 \\ 7 & 8 & 9 \end{pmatrix}$ ii. $\begin{pmatrix} 3 & 7 & -1 \\ 2 & 0 & 3 \\ 8 & 4 & 5 \end{pmatrix}$ iii. $\begin{pmatrix} 7 & -1 & 2 & 4 \\ 3 & 0 & 0 & 4 \\ 1 & 8 & 2 & -1 \\ 3 & 3 & 7 & 2 \end{pmatrix}$

Ans:

i. **0** ii. **54** iii. **783**

A SHORTCUT FOR 3 × 3 MATRICES

Using an easy shortcut known as the Sarrus method, the determinant of a 3 × 3 matrix,

$$A = \begin{pmatrix} a_{11} & a_{12} & a_{13} \\ a_{21} & a_{22} & a_{23} \\ a_{31} & a_{32} & a_{33} \end{pmatrix}$$

can be calculated.

i. **First:** Remove the brackets.

$$\begin{matrix} a_{11} & a_{12} & a_{13} \\ a_{21} & a_{22} & a_{23} \\ a_{31} & a_{32} & a_{33} \end{matrix}$$

ii. **Second:** Repeat the first two columns onto the right.

$$\begin{matrix} a_{11} & a_{12} & a_{13} & a_{11} & a_{12} \\ a_{21} & a_{22} & a_{23} & a_{21} & a_{22} \\ a_{31} & a_{32} & a_{33} & a_{31} & a_{32} \end{matrix}$$

iii. **Third:** Draw pointed lines connecting the elements as shown below,

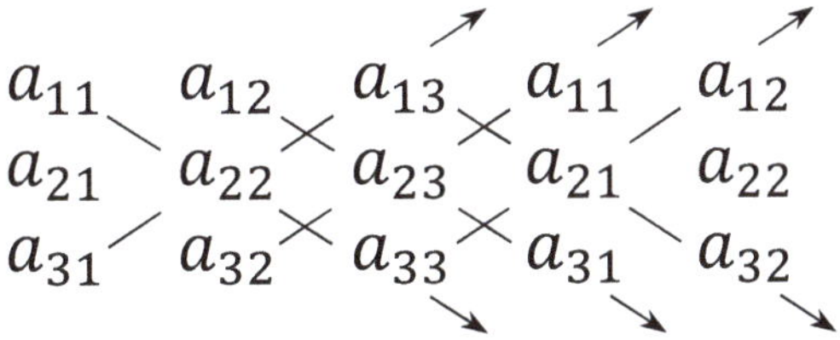

$$\begin{matrix} a_{11} & a_{12} & a_{13} & a_{11} & a_{12} \\ a_{21} & a_{22} & a_{23} & a_{21} & a_{22} \\ a_{31} & a_{32} & a_{33} & a_{31} & a_{32} \end{matrix}$$

iv. **Fourth:** Find the product of the elements along each line and add them.

$$a_{31}a_{22}a_{13} + a_{32}a_{23}a_{11} + a_{33}a_{21}a_{12}$$

$$\begin{matrix} a_{11} & a_{12} & a_{13} & a_{11} & a_{12} \\ a_{21} & a_{22} & a_{23} & a_{21} & a_{22} \\ a_{31} & a_{32} & a_{33} & a_{31} & a_{32} \end{matrix}$$

$$a_{11}a_{22}a_{33} + a_{12}a_{23}a_{31} + a_{13}a_{21}a_{32}$$

v. **Fifth:** To obtain the determinant, subtract the sum above from the sum below. That is,

$$|\mathbf{A}| = (\text{below}) - (\text{above})$$
$$= (a_{11}a_{22}a_{33} + a_{12}a_{23}a_{31} + a_{13}a_{21}a_{32})$$
$$-(a_{31}a_{22}a_{13} + a_{32}a_{23}a_{11} + a_{33}a_{21}a_{12})$$

Example 19: Find the determinant of the 3 × 3 matrix,

$$\begin{pmatrix} 2 & 3 & -1 \\ 0 & 2 & 1 \\ 3 & 4 & 8 \end{pmatrix}$$

Solution: Removing the brackets and repeating the first two columns to the right,

$$\begin{matrix} 2 & 3 & -1 & 2 & 3 \\ 0 & 2 & 1 & 0 & 2 \\ 3 & 4 & 8 & 3 & 4 \end{matrix}$$

Drawing pointed lines,

$$\begin{matrix} 2 & 3 & -1 & 2 & 3 \\ 0 & 2 & 1 & 0 & 2 \\ 3 & 4 & 8 & 3 & 4 \end{matrix}$$

Finding the products of the elements along each line and adding them,

$$(3)(2)(-1) + (4)(1)(2) + (8)(0)(3) = 2$$

$$\begin{matrix} 2 & 3 & -1 & 2 & 3 \\ 0 & 2 & 1 & 0 & 2 \\ 3 & 4 & 8 & 3 & 4 \end{matrix}$$

$$(2)(2)(8) + (3)(1)(3) + (-1)(0)(4) = 41$$

$$|A| = (\text{below}) - (\text{above})$$
$$= (41) - (2)$$
$$= 39$$

Problem 42: Find the determinants using the Sarrus method,

i. $\begin{pmatrix} 1 & 2 & 3 \\ 4 & 5 & 6 \\ 7 & 8 & 9 \end{pmatrix}$ ii. $\begin{pmatrix} 3 & 7 & -1 \\ 2 & 0 & 3 \\ 8 & 4 & 5 \end{pmatrix}$ iii. $\begin{pmatrix} 1 & 1 & -1 \\ 1 & 1 & 1 \\ 1 & -1 & 0 \end{pmatrix}$

Ans:

i. 0 ii. 54 iii. 4

The determinant of a square matrix can also be used to calculate inverses and solve linear equations, which will be the topics for our next chapter.

But for now, let's wrap it up.

Meditate, and recall everything we have studied, and make flash cards for future revisions.

Day Seven

Session One

Cofactor matrix

A shortcut for 3×3 matrices

Adjoint

Session Two

Inverse using determinant

Session Three

Cramer's rule

Every task, goal, race and year comes to an end. Therefore, make a habit to always finish strong.

- Gary Ryan

Before beginning this session, take time to meditate, and recall everything we have studied so far.

If you have flash cards, go through them thoroughly. Today, we will use determinants to find inverses and to solve linear systems.

SESSION ONE

COFACTOR MATRIX

Consider the matrix,

$$A = \begin{pmatrix} a_{11} & a_{12} & a_{13} \\ a_{21} & a_{22} & a_{23} \\ a_{31} & a_{32} & a_{33} \end{pmatrix}$$

with cofactors,

$$A_{ij} = (-1)^{i+j} M_{ij}$$

where, M_{ij} is the minor corresponding to the element a_{ij}.

Then the matrix formed by the cofactors,

$$\begin{pmatrix} A_{11} & A_{12} & A_{13} \\ A_{21} & A_{22} & A_{23} \\ A_{31} & A_{32} & A_{33} \end{pmatrix}$$

Is called the **cofactor matrix** of **A**, written as Cof(**A**) or $\mathbf{A}^{c}$

Example 20: Find the cofactor matrix of,

$$\begin{pmatrix} 2 & 3 & -1 \\ 0 & 2 & 1 \\ 3 & 4 & 8 \end{pmatrix}$$

Solution: First calculate the minors,

$$M_{11} = \begin{vmatrix} 2 & 1 \\ 4 & 8 \end{vmatrix} = 12 \ ; \ M_{12} = \begin{vmatrix} 0 & 1 \\ 3 & 8 \end{vmatrix} = -3 \ ; \ M_{13} = \begin{vmatrix} 0 & 2 \\ 3 & 4 \end{vmatrix} = -6$$

$$M_{21} = \begin{vmatrix} 3 & -1 \\ 4 & 8 \end{vmatrix} = 28 \ ; \ M_{22} = \begin{vmatrix} 2 & -1 \\ 3 & 8 \end{vmatrix} = 19 \ ; \ M_{23} = \begin{vmatrix} 2 & 3 \\ 3 & 4 \end{vmatrix} = -1$$

$$M_{31} = \begin{vmatrix} 3 & -1 \\ 2 & 1 \end{vmatrix} = 5 \ ; \ M_{32} = \begin{vmatrix} 2 & -1 \\ 0 & 1 \end{vmatrix} = 2 \ ; \ M_{33} = \begin{vmatrix} 2 & 3 \\ 0 & 2 \end{vmatrix} = 4$$

Then find the cofactors,

$$A_{11} = (-1)^{1+1}M_{11} = 12\ ;\ A_{12} = (-1)^{1+2}M_{12} = 3\ ;\ A_{13} = (-1)^{1+3}M_{13} = -6$$

$$A_{21} = (-1)^{2+1}M_{21} = -28\ ;\ A_{22} = (-1)^{2+2}M_{22} = 19\ ;\ A_{23} = (-1)^{2+3}M_{23} = 1$$

$$A_{31} = (-1)^{3+1}M_{31} = 5\ ;\ A_{32} = (-1)^{3+2}M_{32} = -2\ ;\ A_{33} = (-1)^{3+3}M_{33} = 4$$

Using the cofactors, we can write the cofactor matrix as,

$$\text{Cof}(A) = \begin{pmatrix} 12 & 3 & -6 \\ -28 & 19 & 1 \\ 5 & -2 & 4 \end{pmatrix}$$

Problem 43: Find the cofactor matrices of,

i. $A = \begin{pmatrix} 1 & 2 & 3 \\ 4 & 5 & 6 \\ 7 & 8 & 9 \end{pmatrix}$ ii. $B = \begin{pmatrix} 3 & 7 & -1 \\ 2 & 0 & 3 \\ 8 & 4 & 5 \end{pmatrix}$

Ans:

i. $\text{cof}(A) = \begin{pmatrix} -3 & 6 & -3 \\ 6 & -12 & 6 \\ -3 & 6 & -3 \end{pmatrix}$ ii. $\text{cof}(B) = \begin{pmatrix} -12 & 14 & 8 \\ -39 & 23 & 44 \\ 21 & -11 & -14 \end{pmatrix}$

AN EASY SHORTCUT FOR 3 × 3 MATRICES

Using an easy shortcut, the cofactor matrix of a 3 × 3 matrix,

$$A = \begin{pmatrix} a_{11} & a_{12} & a_{13} \\ a_{21} & a_{22} & a_{23} \\ a_{31} & a_{32} & a_{33} \end{pmatrix}$$

can be easily calculated.

i. **First:** Repeat the matrix **A** to form a 6×6 matrix.

$$\begin{matrix} a_{11} & a_{12} & a_{13} & a_{11} & a_{12} & a_{13} \\ a_{21} & a_{22} & a_{23} & a_{21} & a_{22} & a_{23} \\ a_{31} & a_{32} & a_{33} & a_{31} & a_{32} & a_{33} \\ a_{11} & a_{12} & a_{13} & a_{11} & a_{12} & a_{13} \\ a_{21} & a_{22} & a_{23} & a_{21} & a_{22} & a_{23} \\ a_{31} & a_{32} & a_{33} & a_{31} & a_{32} & a_{33} \end{matrix}$$

ii. **Second:** Strike out the border elements.

$$\begin{matrix} a_{11} & a_{12} & a_{13} & a_{11} & a_{12} & a_{13} \\ a_{21} & a_{22} & a_{23} & a_{21} & a_{22} & a_{23} \\ a_{31} & a_{32} & a_{33} & a_{31} & a_{32} & a_{33} \\ a_{11} & a_{12} & a_{13} & a_{11} & a_{12} & a_{13} \\ a_{21} & a_{22} & a_{23} & a_{21} & a_{22} & a_{23} \\ a_{31} & a_{32} & a_{33} & a_{31} & a_{32} & a_{33} \end{matrix}$$

iii. **Third:** Find the determinants indicated by the arrows to obtain the Cofactor matrix.

$$\begin{matrix} a_{11} & a_{12} & a_{13} & a_{11} & a_{12} & a_{13} \\ a_{21} & a_{22} & a_{23} & a_{21} & a_{22} & a_{23} \\ a_{31} & a_{32} & a_{33} & a_{31} & a_{32} & a_{33} \\ a_{11} & a_{12} & a_{13} & a_{11} & a_{12} & a_{13} \\ a_{21} & a_{22} & a_{23} & a_{21} & a_{22} & a_{23} \\ a_{31} & a_{32} & a_{33} & a_{31} & a_{32} & a_{33} \end{matrix}$$

$$\text{Cof}(\mathbf{A}) = \begin{pmatrix} a_{22}a_{33} - a_{23}a_{32} & a_{23}a_{31} - a_{21}a_{33} & a_{21}a_{32} - a_{22}a_{31} \\ a_{32}a_{13} - a_{33}a_{12} & a_{33}a_{11} - a_{31}a_{13} & a_{31}a_{12} - a_{32}a_{11} \\ a_{12}a_{23} - a_{13}a_{22} & a_{13}a_{21} - a_{11}a_{23} & a_{11}a_{22} - a_{12}a_{21} \end{pmatrix}$$

Example 21: Find the cofactor matrix of,

$$\begin{pmatrix} 2 & 3 & -1 \\ 0 & 2 & 1 \\ 3 & 4 & 8 \end{pmatrix}$$

Solution: Forming the 6×6 matrix, striking the border elements and finding the determinants,

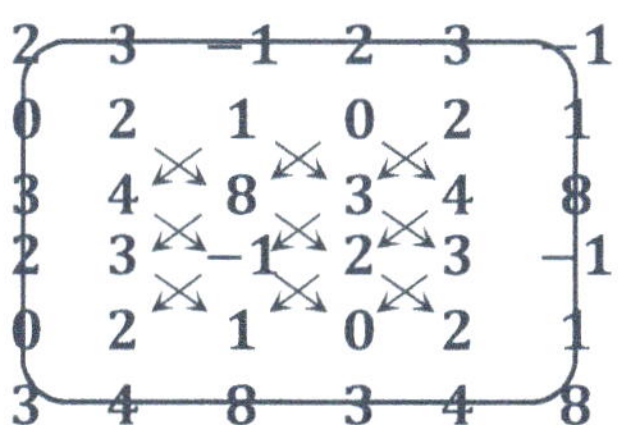

$$\begin{matrix} 2 & 3 & -1 & 2 & 3 & -1 \\ 0 & 2 & 1 & 0 & 2 & 1 \\ 3 & 4 & 8 & 3 & 4 & 8 \\ 2 & 3 & -1 & 2 & 3 & -1 \\ 0 & 2 & 1 & 0 & 2 & 1 \\ 3 & 4 & 8 & 3 & 4 & 8 \end{matrix}$$

$$\text{Cof}(A) = \begin{pmatrix} (2)(8)-(1)(4) & (1)(3)-(0)(8) & (0)(4)-(2)(3) \\ (4)(-1)-(8)(3) & (8)(2)-(3)(-1) & (3)(3)-(4)(2) \\ (3)(1)-(-1)(2) & (-1)(0)-(2)(1) & (2)(2)-(3)(0) \end{pmatrix}$$

$$\text{Cof}(A) = \begin{pmatrix} 12 & 3 & -6 \\ -28 & 19 & 1 \\ 5 & -2 & 4 \end{pmatrix}$$

ADJOINT

Adjoint is the **transpose of the cofactor matrix**. That is, if

$$\text{Cof}(\mathbf{A}) = \begin{pmatrix} A_{11} & A_{12} & A_{13} \\ A_{21} & A_{22} & A_{23} \\ A_{31} & A_{32} & A_{33} \end{pmatrix}$$

Then,

$$\text{Adj}(\mathbf{A}) = \left(\text{Cof}(\mathbf{A})\right)^{T} = \begin{pmatrix} A_{11} & A_{21} & A_{31} \\ A_{12} & A_{22} & A_{32} \\ A_{13} & A_{23} & A_{33} \end{pmatrix}$$

Problem 44: Find the adjoints of the following matrices,

i. $A = \begin{pmatrix} 3 & -1 & 0 \\ 0 & 1 & 2 \\ 4 & 8 & 5 \end{pmatrix}$ ii. $B = \begin{pmatrix} 1 & 7 & 6 \\ 8 & -1 & -1 \\ 2 & 7 & 8 \end{pmatrix}$

Ans:

i. $\text{Adj}(A) = \begin{pmatrix} -11 & 5 & -2 \\ 8 & 15 & -6 \\ -4 & -28 & 3 \end{pmatrix}$ ii. $\text{Adj}(B) = \begin{pmatrix} -1 & -14 & -1 \\ -66 & -4 & 49 \\ 58 & 7 & -57 \end{pmatrix}$

The shortcut method is easier to use and simpler to execute. So, if the original method is hard, try to memorize the shortcut.

In the next session, we will learn a new method for finding inverses in which adjoints and determinants play a crucial part.

SESSION TWO

INVERSES USING DETERMINANT

If $\mathbf{A}$ is a square matrix with a determinant $|\mathbf{A}|$, and an adjoint $\mathrm{Adj}(\mathbf{A})$, then its inverse is,

$$\mathbf{A}^{-1} = \frac{1}{|\mathbf{A}|}\mathrm{Adj}(\mathbf{A})$$

See the example,

Example 22: Find the inverse of,

$$\begin{pmatrix} 1 & 2 & -1 \\ 0 & 1 & 1 \\ 2 & 3 & 1 \end{pmatrix}$$

Solution: First find the cofactor matrix (for simplicity let's use the shortcut),

$$\begin{array}{cccccc} 2 & 3 & -1 & 2 & 3 & -1 \\ 0 & 2 & 1 & 0 & 2 & 1 \\ 3 & 4 & 8 & 3 & 4 & 8 \\ 2 & 3 & -1 & 2 & 3 & -1 \\ 0 & 2 & 1 & 0 & 2 & 1 \\ 3 & 4 & 8 & 3 & 4 & 8 \end{array}$$

$$\mathbf{Cof(A)} = \begin{pmatrix} -2 & 2 & -2 \\ -5 & 3 & 1 \\ 3 & -1 & 1 \end{pmatrix}$$

From the cofactor matrix, we get,

$$\mathrm{Adj}(\mathbf{A}) = \left(\mathrm{Cof}(\mathbf{A})\right)^{\mathrm{T}} = \begin{pmatrix} -2 & 2 & -2 \\ -5 & 3 & 1 \\ 3 & -1 & 1 \end{pmatrix}^{\mathrm{T}} = \begin{pmatrix} -2 & -5 & 3 \\ 2 & 3 & -1 \\ -2 & 1 & 1 \end{pmatrix}$$

Also, from the cofactor matrix, we get,

$$\mathbf{A_{11} = -2 \;;\; A_{12} = 2 \;;\; A_{13} = -2}$$

Therefore,

$$|\mathrm{A}| = \begin{vmatrix} 1 & 2 & -1 \\ 0 & 1 & 1 \\ 2 & 3 & 1 \end{vmatrix} = a_{11}\mathrm{A}_{11} + a_{12}\mathrm{A}_{12} + a_{13}\mathrm{A}_{13}$$

$$= (1)(-2) + (2)(2) + (-1)(-2)$$

$$= 4$$

From Adj(A) and |A| we get the inverse,

$$A^{-1} = \frac{1}{|A|}\text{Adj}(A) = \frac{1}{4}\begin{pmatrix} -2 & -5 & 3 \\ 2 & 3 & -1 \\ -2 & 1 & 1 \end{pmatrix}$$

$$= \begin{pmatrix} \left(\frac{1}{4}\right)(-2) & \left(\frac{1}{4}\right)(-5) & \left(\frac{1}{4}\right)(3) \\ \left(\frac{1}{4}\right)(2) & \left(\frac{1}{4}\right)(3) & \left(\frac{1}{4}\right)(-1) \\ \left(\frac{1}{4}\right)(-2) & \left(\frac{1}{4}\right)(1) & \left(\frac{1}{4}\right)(1) \end{pmatrix}$$

$$= \begin{pmatrix} -1/2 & -5/4 & 3/4 \\ 1/2 & 3/4 & -1/4 \\ -1/2 & 1/4 & 1/4 \end{pmatrix}$$

(See example 6 and problem 26)

Problem 45: Find the inverses of the following matrices,

i. $\begin{pmatrix} 2 & 0 \\ 1 & 1 \end{pmatrix}$ ii. $\begin{pmatrix} 1 & 0 & 2 \\ 2 & -1 & 3 \\ 4 & 1 & 8 \end{pmatrix}$ iii. $\begin{pmatrix} 2 & 3 \\ 2 & 1 \end{pmatrix}$

iv. $\begin{pmatrix} 9 & 6 \\ 3 & 2 \end{pmatrix}$ v. $\begin{pmatrix} 3 & 0 & 0 \\ -1 & 2 & 0 \\ 1 & 1 & 0 \end{pmatrix}$

Ans:

i. $\begin{pmatrix} 1/2 & 0 \\ -1/2 & 1 \end{pmatrix}$ ii. $\begin{pmatrix} -11 & 2 & 2 \\ -4 & 0 & 1 \\ 6 & -1 & -1 \end{pmatrix}$ iii. $\begin{pmatrix} -1/4 & 3/4 \\ 1/2 & -1/2 \end{pmatrix}$

iv. No Inverse v. No Inverse

The determinant of the matrices **iv** and **v** in the above problem were equal to zero. Since division by zero is not defined, reciprocals of the determinants cannot be found. That is, since,

$$|A| = 0 \;\; ; \;\; \frac{1}{|A|} \text{ is not defined (See Day Six} - \text{ Session One)}$$

showing that the matrices are singular.

This further validates the importance of determinants. They not only tell us about the consistency of a system, but also about the invertibility of the matrix itself.

To summarize,

If the determinant of a square matrix is zero, then the matrix is singular.

The applications of determinants extend ever further. Determinants provide an easy way to solve linear systems called the Cramer's rule, named after Gabriel Cramer (1704-1752). We will discuss the Cramer's rule in the next session, but for now, let's take a break.

SESSION THREE

CRAMER'S RULE

Cramer's rule is a surprisingly simple method for solving linear systems. To demonstrate the method, consider the system,

$$\left.\begin{aligned} a_{11}x_1 + a_{12}x_2 = b_1 \\ a_{21}x_1 + a_{22}x_2 = b_2 \end{aligned}\right\}$$

In matrix form,

$$\begin{pmatrix} a_{11} & a_{12} \\ a_{21} & a_{22} \end{pmatrix}\begin{pmatrix} x_1 \\ x_2 \end{pmatrix} = \begin{pmatrix} b_1 \\ b_2 \end{pmatrix}$$

i. **First:** Find the determinant of the coefficient matrix.

$$|\mathbf{A}| = \begin{vmatrix} a_{11} & a_{12} \\ a_{21} & a_{22} \end{vmatrix}$$

If $|\mathbf{A}| = 0$, stop. The rule cannot be applied.

If $|\mathbf{A}| \neq 0$, then,

iv. **Second:** Replace the first column of the coefficient matrix by the RHS matrix. That is,

$$\text{replace, } \begin{pmatrix} a_{11} \\ a_{21} \end{pmatrix} \text{ on } \begin{pmatrix} a_{11} & a_{12} \\ a_{21} & a_{22} \end{pmatrix} \text{ by } \begin{pmatrix} b_1 \\ b_2 \end{pmatrix} \text{ to get } \begin{pmatrix} b_1 & a_{12} \\ b_2 & a_{22} \end{pmatrix}$$

v. **Third:** Find its determinant and call it $|\hat{\mathrm{x}}_1|$. That is,

$$|\hat{\mathrm{x}}_1| = \begin{vmatrix} b_1 & a_{12} \\ b_2 & a_{22} \end{vmatrix}$$

vi. **Fourth:** Similarly, find $|\hat{x}_2|$ by replacing the second column of the coefficient matrix by the RHS matrix and finding its determinant. That is,

$$|\hat{x}_2| = \begin{vmatrix} a_{11} & b_1 \\ a_{21} & b_2 \end{vmatrix}$$

vii. **Fifth:** Using $|\mathbf{A}|$, $|\hat{x}_1|$ and $|\hat{x}_2|$ we obtain the solutions,

$$x_1 = \frac{|\hat{x}_1|}{|\mathbf{A}|} = \frac{\begin{vmatrix} b_1 & a_{12} \\ b_2 & a_{22} \end{vmatrix}}{\begin{vmatrix} a_{11} & a_{12} \\ a_{21} & a_{22} \end{vmatrix}}$$

and,

$$x_2 = \frac{|\hat{x}_2|}{|\mathbf{A}|} = \frac{\begin{vmatrix} a_{11} & b_1 \\ a_{21} & b_2 \end{vmatrix}}{\begin{vmatrix} a_{11} & a_{12} \\ a_{21} & a_{22} \end{vmatrix}}$$

See the example,

Example 23: Solve the following system using Cramer's rule,

$$\left.\begin{aligned} 2x_1 + x_2 - x_3 &= 2 \\ 2x_1 + x_3 &= 3 \\ x_1 - x_2 &= 0 \end{aligned}\right\}$$

Solution: In matrix form, the system becomes,

$$\begin{pmatrix} 2 & 1 & -1 \\ 2 & 0 & 1 \\ 1 & -1 & 0 \end{pmatrix}\begin{pmatrix} x_1 \\ x_2 \\ x_3 \end{pmatrix} = \begin{pmatrix} 2 \\ 3 \\ 0 \end{pmatrix}$$

First, let us find the determinant of the coefficient matrix,

$$\mathrm{A} = \begin{pmatrix} 2 & 1 & -1 \\ 2 & 0 & 1 \\ 1 & -1 & 0 \end{pmatrix}$$

Let's expand along the second row, since it contains a zero. That is,

$$|\mathrm{A}| = \begin{vmatrix} 2 & 1 & -1 \\ 2 & 0 & 1 \\ 1 & -1 & 0 \end{vmatrix} = a_{21}\mathrm{A}_{21} + a_{22}\mathrm{A}_{22} + a_{23}\mathrm{A}_{23}$$
$$= (2)\mathrm{A}_{21} + (0)\mathrm{A}_{22} + (1)\mathrm{A}_{23}$$
$$= (2)\mathrm{A}_{21} + (1)\mathrm{A}_{23}$$

We have,

$$\mathrm{M}_{21} = \begin{vmatrix} 1 & -1 \\ -1 & 0 \end{vmatrix} = -1\ ;\ \text{therefore}, \mathrm{A}_{21} = (-1)^{2+1}\mathrm{M}_{21} = 1$$

$$M_{23} = \begin{vmatrix} 2 & 1 \\ 1 & -1 \end{vmatrix} = -3 \; ; \; \text{therefore}, A_{23} = (-1)^{2+3} M_{23} = 3$$

Thus, we get,

$$|A| = (2)(1) + (1)(3) = 5$$

Now, replacing the first column of the coefficient matrix by the RHS matrix, and finding its determinant. That is,

$$|\hat{x}_1| = \begin{vmatrix} 2 & 1 & -1 \\ 3 & 0 & 1 \\ 0 & -1 & 0 \end{vmatrix} = (0)A_{31} + (-1)A_{32} + (0)A_{33}$$

$$= (-1)A_{32}$$

We have,

$$M_{32} = \begin{vmatrix} 2 & -1 \\ 3 & 1 \end{vmatrix} = 5 \; ; \; \text{therefore}, A_{32} = (-1)^{3+2} M_{32} = -5$$

Thus, we get,

$$|\hat{x}_1| = (-1)(-5) = 5$$

Replacing the second column of the coefficient matrix by the RHS matrix, and finding its determinant. That is,

$$|\hat{x}_2| = \begin{vmatrix} 2 & 2 & -1 \\ 2 & 3 & 1 \\ 1 & 0 & 0 \end{vmatrix} = (1)A_{31} + (0)A_{32} + (0)A_{33}$$

$$= (1)A_{31}$$

We have,

$$M_{31} = \begin{vmatrix} 2 & -1 \\ 3 & 1 \end{vmatrix} = 5 \; ; \; \text{therefore}, A_{31} = (-1)^{3+1} M_{31} = 5$$

Thus, we get,

$$|\hat{x}_1| = (1)(5) = 5$$

Replacing the third column of the coefficient matrix by the RHS matrix, and finding its determinant. That is,

$$|\hat{x}_3| = \begin{vmatrix} 2 & 1 & 2 \\ 2 & 0 & 3 \\ 1 & -1 & 0 \end{vmatrix} = (2)A_{21} + (0)A_{22} + (3)A_{23}$$

$$= (2)A_{21} + (3)A_{23}$$

We have,

$$M_{21} = \begin{vmatrix} 1 & 2 \\ -1 & 0 \end{vmatrix} = 2 \text{ ; therefore, } A_{21} = (-1)^{2+1}M_{21} = -2$$

$$M_{23} = \begin{vmatrix} 2 & 1 \\ 1 & -1 \end{vmatrix} = -3 \text{ ; therefore, } A_{23} = (-1)^{2+3}M_{23} = 3$$

Thus, we get,

$$|\hat{x}_3| = (2)(-2) + (3)(3) = 5$$

Using, $|A|$, $|\hat{x}_1|$, $|\hat{x}_2|$, $|\hat{x}_3|$, we get, the solutions.

$$x_1 = \frac{|\hat{x}_1|}{|A|} = \frac{5}{5} = 1$$

$$x_2 = \frac{|\hat{x}_2|}{|A|} = \frac{5}{5} = 1$$

$$x_3 = \frac{|\hat{x}_3|}{|A|} = \frac{5}{5} = 1$$

That is,

$$\left.\begin{aligned} x_1 &= 1 \\ x_2 &= 1 \\ x_3 &= 1 \end{aligned}\right\} \text{ or } (1, 1, 1)$$

Notice that the Cramer's rule can be applied only if the determinant of the coefficient matrix is non-zero. Since, the division by a zero determinant is not possible.

Problem 46: Solve the following linear systems using the Cramer's rule.

i. $\left.\begin{aligned} 2x - y &= 1 \\ 3x + 2y &= 12 \end{aligned}\right\}$ ii. $\left.\begin{aligned} x + 2y - z &= 1 \\ 2x + y + 4z &= 2 \\ 3x + 3y + 4z &= 1 \end{aligned}\right\}$ iii. $\left.\begin{aligned} x + y &= -2 \\ x - 2y &= 7 \end{aligned}\right\}$

iv. $\left.\begin{aligned} x + 2y &= 1 \\ 2x + 4y &= 2 \end{aligned}\right\}$ v. $\left.\begin{aligned} 2x + 3y + 5z &= 7 \\ x + 2y + 3z &= 1 \\ 3x + 6y + 9z &= 3 \end{aligned}\right\}$

Ans:

i. $\left.\begin{aligned} x &= 2 \\ y &= 3 \end{aligned}\right\}$ ii. $\left.\begin{aligned} x &= 7 \\ y &= -4 \\ z &= -2 \end{aligned}\right\}$ iii. $\left.\begin{aligned} x &= 1 \\ y &= -3 \end{aligned}\right\}$

iv. **Rule Not Applicable** v. **Rule Not Applicable**

We end our course with the discussion of Cramer's rule. For some, the sessions may seem too big for one week. If so, take your own time. The key is to be persistent. Given enough time and focus, anything is possible.

The scope and applications of matrices extends far beyond a one-week course. From quantum mechanics to machine learning, matrices appear in almost every branch of science and mathematics.

I hope this book provided a tiny yet self-consistent introduction to the more general theory of linear systems called 'Linear algebra'.

For further study, refer:

Linear algebra by, Jim Hefferon,

One more thing,

Before ending the session, remember to meditate and recall everything. If you are satisfied with the sessions, you are good to go. Have a good night sleep and come back tomorrow for solving the 'Final Quiz'. If you are not satisfied, it's ok. Tomorrow, return to the same session.

Don't give up.

Final Quiz

Even the hardest puzzles have a solution.

- Unknown

PART-A

1) Find the orders of the following matrices.

i. (0) ii. $\begin{pmatrix} 0 & 1 & 8 \\ 2 & 7 & 5 \end{pmatrix}$ iii. $\begin{pmatrix} 1 & x & y & 0 \\ 8 & a & b & 1 \\ 0 & 0 & 0 & 0 \end{pmatrix}$ iv. $\begin{pmatrix} 0 \\ 0 \\ 0 \end{pmatrix}$

2) If $\mathbf{A} = \begin{pmatrix} x & y & {}^{-1}/_{\sqrt{3}} & 7 \\ 8 & 9 & 0 & 0 \\ 0 & -1 & 1 & 0 \end{pmatrix}$, find:

i. a_{11} ii. a_{13} iii. a_{23} iv. a_{34}

3) Find the transposes of the following matrices.

i. $\begin{pmatrix} 1 & \sqrt{3} \\ 0 & 1 \end{pmatrix}$ ii. $\begin{pmatrix} 0 & 1 \\ 1 & 0 \end{pmatrix}$ iii. $\begin{pmatrix} 1 & 1 \\ 0 & x \\ 0 & -2 \end{pmatrix}$ iv. $(1 \quad 2 \quad 3)$

4) Which of the following matrices are symmetric?

i. $\begin{pmatrix} 1 & 0 & 0 \\ 0 & 1 & 0 \\ 0 & 0 & -1 \end{pmatrix}$ ii. $\begin{pmatrix} 1 & 2 & -1 \\ 2 & 0 & 1 \\ 1 & 1 & 0 \end{pmatrix}$ iii. $\begin{pmatrix} 1 & -1 \\ 1 & 1 \end{pmatrix}$ iv. $\begin{pmatrix} 0 & 0 \\ 0 & 0 \end{pmatrix}$

5) Which of the following matrices are anti-symmetric?

i. $\begin{pmatrix} 0 & 1 & 2 \\ -1 & 0 & 3 \\ -2 & -3 & 0 \end{pmatrix}$ ii. $\begin{pmatrix} 1 & -2 \\ 2 & 1 \end{pmatrix}$ iii. $\begin{pmatrix} 0 & 1 \\ 1 & 0 \end{pmatrix}$ iv. $\begin{pmatrix} 0 & 0 \\ 0 & 0 \end{pmatrix}$

6) If $\begin{pmatrix} x & {}^{1}/_{y} & -1 \\ 2 & 0 & 3 \\ 1 & 3 & z \end{pmatrix}$ is anti-symmetric, find:

i. x ii. y iii. z

7) If $\mathbf{A} = \begin{pmatrix} 1 & 0 & 5 \\ -2 & 1 & 6 \\ 3 & 2 & 7 \end{pmatrix}$, find $\mathbf{A} + \mathbf{A}^{\mathrm{T}}$ and $\mathbf{A} - \mathbf{A}^{\mathrm{T}}$. Also show that one is symmetric and the other is anti-symmetric.

PART-B

1) If $\mathbf{A} = \begin{pmatrix} 2 & 2 \\ 3 & -4 \end{pmatrix}$; $\mathbf{B} = \begin{pmatrix} 1 & 0 & -1 \\ 2 & -1 & 0 \\ 0 & 0 & 3 \end{pmatrix}$; $\mathbf{C} = \begin{pmatrix} 2 & -1 \\ 3 & 4 \\ 0 & 1 \end{pmatrix}$; $\mathbf{D} = \begin{pmatrix} 1 & -1 \\ 2 & 1 \end{pmatrix}$

and $\mathbf{E} = \begin{pmatrix} 7 & 2 & 3 \\ -11 & 0 & -2 \\ 8 & 9 & -2 \end{pmatrix}$, find:

i. $\mathbf{A+D}$ ii. $\mathbf{2B}$ iii. $(\mathbf{B^T})(\mathbf{E})$ iv. $\mathbf{A-2D}$

v. $\mathbf{BE}$ vi. $\mathbf{CB}$ vii. $\mathbf{BC}$ viii. $\mathbf{CA+C}$

2) If $\mathbf{A} = \begin{pmatrix} 0 \\ -1 \\ 2 \end{pmatrix}$ and $\mathbf{B} = (3 \quad 7 \quad 8)$. Find:

i. $\mathbf{AB}$ ii. $\mathbf{BA}$ iii. $\mathbf{A^TB}$ iv. $\mathbf{B^TA^T}$ v. $\mathbf{AB^T}$

3) If $\mathbf{A} = \begin{pmatrix} 1 & 2 & 2 \\ 2 & 1 & 2 \\ 2 & 2 & 1 \end{pmatrix}$, show that:

$$\mathbf{A^2 - 4A - 5I = 0}.$$

4) If $\mathbf{A} = \begin{pmatrix} 3 & -1 & 7 \\ 8 & 0 & 3 \\ 1 & 4 & 9 \end{pmatrix}$. Find the minors:

i. M_{11} ii. M_{31} iii. M_{23}

and the cofactors:

i. A_{21} ii. A_{13} iii. A_{22}

5) Find the determinants of the following matrices.

i. $\begin{pmatrix} 1 & 0 \\ -1 & 1 \end{pmatrix}$ ii. $\begin{pmatrix} 2 & 1 & 2 \\ 7 & -1 & -3 \\ 7 & 8 & 4 \end{pmatrix}$ iii. $\begin{pmatrix} 1 & -1 \\ 2 & -1 \\ 3 & 2 \end{pmatrix}$

iv. $\begin{pmatrix} 0 & 1 & 2 & 3 \\ 7 & 9 & 11 & 2 \\ 8 & 4 & 2 & 3 \\ 1 & -1 & 2 & -1 \end{pmatrix}$

6) Solve for 'x' if

$$\begin{vmatrix} 3 & 1 & 9 \\ 2x & 2 & 6 \\ x^2 & 3 & 3 \end{vmatrix} = 0$$

7) Solve for 'x' if:

$$\begin{vmatrix} 2 & 1 & x \\ 3 & -1 & 2 \\ 1 & 1 & 6 \end{vmatrix} = \begin{vmatrix} 4 & x \\ 3 & 2 \end{vmatrix}$$

PART-C

1) Execute the following row operations on the matrix,

$$\begin{pmatrix} 7 & 8 & 0 \\ 9 & 6 & 4 \\ 2 & -1 & -3 \end{pmatrix}$$

i. $R_1 \leftrightarrow R_2$ ii. $R_2 \rightarrow -\frac{7}{3} R_2$

iii. $R_3 \rightarrow 2R_1 + R_3$ iv. $R_2 \rightarrow -\frac{2}{3} R_1 + R_2$

2) Which of the following matrices are in Row-Echelon form:

i. $\begin{pmatrix} 1 & 0 & 0 \\ 0 & 1 & 0 \\ 0 & 0 & 1 \end{pmatrix}$ ii. $\begin{pmatrix} 1 & 0 & 0 \\ 0 & -1 & 0 \\ 0 & 0 & 1 \end{pmatrix}$ iii. $\begin{pmatrix} 1 & 2 & 3 \\ 0 & 0 & 0 \\ 0 & 0 & 0 \end{pmatrix}$

iv. $\begin{pmatrix} 1 & 4 & 8 \\ 0 & 1 & 3 \\ 0 & 0 & 0 \end{pmatrix}$

3) Reduce the following matrices into Row-Echelon form using elementary row-operations. Also find their ranks.

i. $\begin{pmatrix} 3 & 8 & 0 \\ 6 & 6 & 4 \\ 0 & -1 & -3 \end{pmatrix}$ ii. $\begin{pmatrix} 2 & -1 \\ 0 & -8 \end{pmatrix}$ iii. $\begin{pmatrix} 0 & 1 \\ 2 & 3 \\ 8 & -1 \end{pmatrix}$

4) If a matrix **A** has rank 2, find the rank of $\mathbf{A}^2$.

(Hint: find the answer by using a sample matrix)

5) Which of the following pairs of matrices are inverses of each other?

i. $\begin{pmatrix} 1 & 2 \\ 3 & 4 \end{pmatrix}, \begin{pmatrix} -2 & 1/2 \\ -3/2 & 1/2 \end{pmatrix}$ ii. $\begin{pmatrix} 5 & 8 & 0 \\ 0 & 2 & 1 \\ 4 & 3 & 1 \end{pmatrix}, \begin{pmatrix} 5 & -11 & -6 \\ -4 & 9 & 5 \\ 8 & -17 & -10 \end{pmatrix}$

iii. $\begin{pmatrix} 1 & 2 \\ 3 & 4 \end{pmatrix}, \begin{pmatrix} -2 & 1 \\ 3/2 & -1/2 \end{pmatrix}$ iv. $\begin{pmatrix} -8 & 5 & -4 \\ 9 & -1 & 0 \\ 0 & -2 & 0 \end{pmatrix}, \begin{pmatrix} -3/2 & -1/2 & -1/2 \\ 8/2 & -5/2 & 2/4 \\ 7/2 & 3/2 & 9/2 \end{pmatrix}$

6. Find the inverse of the following matrices using elementary row operations.

i. $\begin{pmatrix} 1 & 1 \\ 1 & 1 \end{pmatrix}$ ii. $\begin{pmatrix} 1 & 2 \\ 3 & 4 \end{pmatrix}$ iii. $\begin{pmatrix} 1 & 2 & 3 \\ 4 & 5 & 6 \\ 7 & 8 & 9 \end{pmatrix}$ iv. $\begin{pmatrix} 5 & 8 & 1 \\ 0 & 2 & 1 \\ 4 & 3 & -1 \end{pmatrix}$

7. What is the role played by rank in finding inverses?

PART-D

1. Solve the following linear systems using inverse method.

   i. $\left.\begin{array}{r} 3x+5y=6 \\ x+y=2 \end{array}\right\}$ ii. $\left.\begin{array}{r} y-z=3 \\ 3x+2y-z=4 \\ 3x-2y+z=1 \end{array}\right\}$ iii. $\left.\begin{array}{r} x+y-z=0 \\ 2x-3y+z=1 \\ 2x+y+2z=7 \end{array}\right\}$

2. What is the significance of rank in gaussian elimination?
3. Determine whether the following systems are consistent or not:

   i. $\left.\begin{array}{r} x+y=-2 \\ x-2y=7 \end{array}\right\}$ ii. $\left.\begin{array}{r} 5x-2y=-7 \\ -5x+2y=2 \end{array}\right\}$ iii. $\left.\begin{array}{r} x+2y-z=1 \\ 2x+y+4z=2 \\ 3x+3y+4z=1 \end{array}\right\}$

4. Solve the following systems using gaussian elimination:

   i. $\left.\begin{array}{r} 3x+2y=12 \\ 3x+2y=6 \end{array}\right\}$ ii. $\left.\begin{array}{r} x+y+z=2 \\ 6x-4y+5z=31 \\ 5x+2y+2z=13 \end{array}\right\}$

   iii. $\left.\begin{array}{r} 2x-5y+5z=17 \\ x-3y+z=6 \\ x-2y+3z=9 \end{array}\right\}$ iv. $\left.\begin{array}{r} 3x-y-z=0 \\ x+y+2z=0 \\ 5x+y+3z=0 \end{array}\right\}$

5. What is the connection between inverse and determinant?
6. Find the inverse of the following matrices by the determinant method:

   i. $\begin{pmatrix} 1 & 1 \\ 1 & 1 \end{pmatrix}$ ii. $\begin{pmatrix} 1 & 2 \\ 3 & 4 \end{pmatrix}$ iii. $\begin{pmatrix} 1 & 2 & 3 \\ 4 & 5 & 6 \\ 7 & 8 & 9 \end{pmatrix}$ iv. $\begin{pmatrix} 5 & 8 & 1 \\ 0 & 2 & 1 \\ 4 & 3 & -1 \end{pmatrix}$

7. Solve the following linear systems using Cramer's rule:

   i. $\left.\begin{array}{r} 3x+2y=12 \\ 3x+2y=6 \end{array}\right\}$ ii. $\left.\begin{array}{r} x+y+z=2 \\ 6x-4y+5z=31 \\ 5x+2y+2z=13 \end{array}\right\}$

   iii. $\left.\begin{array}{r} 2x-5y+5z=17 \\ x-3y+z=6 \\ x-2y+3z=9 \end{array}\right\}$ iv. $\left.\begin{array}{r} 3x-y-z=0 \\ x+y+2z=0 \\ 5x+y+3z=0 \end{array}\right\}$

The journey of a thousand miles begins with one step.

- Lao Tzu

NOTES

NOTES

www.ingramcontent.com/pod-product-compliance
Ingram Content Group UK Ltd.
Pitfield, Milton Keynes, MK11 3LW, UK
UKHW050146280726
14058UKWH00007B/861